AF740633

Springer Aerospace Technology

Series Editors

Sergio De Rosa, Department of Industrial Engineering, University of Naples Federico II, Napoli, Italy

Yao Zheng, School of Aeronautics and Astronautics, Zhejiang University, Hangzhou, China

Elena Popova, Air Navigation Bridge Russia, Chelyabinsk, Russia

The series explores the technology and the science related to the aircraft and spacecraft including concept, design, assembly, control and maintenance. The topics cover aircraft, missiles, space vehicles, aircraft engines and propulsion units. The volumes of the series present the fundamentals, the applications and the advances in all the fields related to aerospace engineering, including:

- structural analysis,
- aerodynamics,
- aeroelasticity,
- aeroacoustics,
- flight mechanics and dynamics,
- orbital maneuvers,
- avionics,
- systems design,
- materials technology,
- launch technology,
- payload and satellite technology,
- space industry, medicine and biology.

The series' scope includes monographs, professional books, advanced textbooks, as well as selected contributions from specialized conferences and workshops.

The volumes of the series are single-blind peer-reviewed.

To submit a proposal or request further information, please contact: Mr. Pierpaolo Riva at pierpaolo.riva@springer.com (Europe and Americas) Mr. Mengchu Huang at mengchu.huang@springer.com (China)

The series is indexed in Scopus and Compendex

Dilmurat M. Azimov · Robert H. Bishop

Apollo-Class Targeting and Guidance for Lunar Powered Descent and Precision Landing

Springer

Dilmurat M. Azimov
Department of Mechanical Engineering
University of Hawai'i at Manoa
Honolulu, Hawai'i, USA

Robert H. Bishop
Department of Aerospace Engineering
Texas A&M University
College Station, Texas, USA

ISSN 1869-1730 ISSN 1869-1749 (electronic)
Springer Aerospace Technology
ISBN 978-3-031-91087-6 ISBN 978-3-031-91088-3 (eBook)
https://doi.org/10.1007/978-3-031-91088-3

© The Editor(s) (if applicable) and The Author(s), under exclusive license to Springer Nature Switzerland AG 2026

This work is subject to copyright. All rights are solely and exclusively licensed by the Publisher, whether the whole or part of the material is concerned, specifically the rights of translation, reprinting, reuse of illustrations, recitation, broadcasting, reproduction on microfilms or in any other physical way, and transmission or information storage and retrieval, electronic adaptation, computer software, or by similar or dissimilar methodology now known or hereafter developed.
The use of general descriptive names, registered names, trademarks, service marks, etc. in this publication does not imply, even in the absence of a specific statement, that such names are exempt from the relevant protective laws and regulations and therefore free for general use.
The publisher, the authors and the editors are safe to assume that the advice and information in this book are believed to be true and accurate at the date of publication. Neither the publisher nor the authors or the editors give a warranty, expressed or implied, with respect to the material contained herein or for any errors or omissions that may have been made. The publisher remains neutral with regard to jurisdictional claims in published maps and institutional affiliations.

This Springer imprint is published by the registered company Springer Nature Switzerland AG
The registered company address is: Gewerbestrasse 11, 6330 Cham, Switzerland

If disposing of this product, please recycle the paper.

To
Mamura, Komola, Feruza, Dilshod and Ra'no

To
Lynda

Foreword

Over the last two decades, research and development of guidance, navigation and control systems have been intensified along with NASA's strategy and efforts toward the Moon and Mars exploration missions involving private industries. The critical goals of these efforts, such as returning to the Moon and sending humans to Mars have resonated among scientific and engineering communities across the globe. Understanding the heritage of Apollo missions, especially how the computer programs on-board the spacecraft and Lunar Modules executed to land astronauts on the Moon have been of interest to the previous Mars and Moon landing missions, such as Mars Science Lab and Intuitive Machines' IM-1 and IM-2 missions to the Lunar South Pole. These efforts truly stand on the shoulders of the giants in guidance, navigation and control systems who came before in the Apollo program.

However, the contents of the NASA's Apollo missions-related internal pre-flight and post-flight documents as well as the papers published on these topics and the users guides were not intended for use by today's readers and today's users. In fact, without a clear understanding of these materials, especially Apollo guidance and targeting procedures, it would be very difficult even for a skilled computer engineer to reconstruct the algorithms of the Apollo guidance computer and duplicate Apollo lunar descent guidance performance. It is therefore fortunate that the authors took on the forensic effort to locate, integrate, and translate the available material on Apollo targeting and guidance for a new generation of explorers.

This book provides rigorous and step-by-step explanations and descriptions of the Apollo guidance algorithms, the execution of which allows for duplication of the lunar descent guidance performance. In Part II of the book, the authors modified the Apollo lunar descent guidance algorithms by introducing a new analytical solution for a constant thrust arc of nearly optimal trajectories. This modification replaces the Apollo ground-based targeting programs with on-board and real-time targeting and retargeting programs thereby leveraging the autonomous capabilities of landers. The authors have also shown how the modified targeting programs and coordinate transformations allow for an increased accuracy of landing. Therefore, this book would be beneficial to all students, engineers and scientists working with the problems of design and performance of planetary descent and landing maneuvers. It will also be useful to those working in the design and development of guidance, navigation and control software to leverage a system's autonomy by enhancing its autonomous capabilities. There are many ways to approach powered planetary descent targeting and guidance, but the reader will now find an updated and enhanced version of the proven Apollo approach as an effective starting point for their own implementations and exploration of onboard algorithms and systems analysis.

April 2025

Dr. Timothy Crain
Intuitive Machines
Houston, Texas, USA

Preface

Our aim is to provide a detailed explanation of the powered lunar descent guidance algorithms used during the Apollo missions and how these algorithms can be modified to enhance future planetary landers with autonomous capabilities.

The book is based on years of collaboration between the authors on research involving the derivation of analytical solutions for thrust arcs and the development of guidance laws for different classes of maneuvers. It primarily presents the results of research initially conducted to address the Orion mission to the Moon in the early 2000s and to investigate the problem of powered descent and landing for future Moon missions. This research involved comprehensive studies of many technical documents describing the Apollo missions and their lunar descent guidance performance. Understanding the Apollo guidance computer programs and their execution to ensure safe human landings on the Moon can be very insightful for future successful landing maneuvers. However, the Apollo-era documents, especially those describing the guidance, navigation, and control (GNC) of the landing module, were intended for internal use and not as a user guide for a wide range of students and engineers interested in powered lunar descent and landing guidance.

Part I of the book is dedicated to the reconstruction of the performance of the powered descent and landing maneuver based on the Apollo guidance computer programs. These programs provide algorithms for ground-based targeting using repetitive iterative procedures and on-board guidance for powered descent.

As Henry Kelly, a pioneer of guidance, stated, analytical solutions are the "heart" of guidance. These solutions can reveal the general laws that lead to a clear understanding of the nature of motion for which the guidance and control schemes are executed. Planetary descent and landing guidance have attracted many studies since the late 1990s. Since then, general numerical and analytical guidance methods have been developed. In the absence of analytical solutions, if the problem can be reduced to a more idealized form that allows for such solutions, these solutions can serve as an initial iteration to obtain more accurate solutions for real-world problems.

Analytical mechanics is also finding applications in engineering practice, particularly in flight dynamics and control through Hamiltonian formalism. Studies of controllable systems have shown that methods developed based on this formalism possess a high degree of analyticity, allowing for convenient algorithmization of the process of interest. This book demonstrates how numerical and analytical guidance methods can complement each other, opening greater opportunities for the design and implementation of onboard guidance and control schemes, which are covered in the second part of the book.

Part II focuses on the development of modifications to the Apollo guidance and targeting programs and their execution to demonstrate their utility for future landing missions. The targeting programs are based on analytical solutions to the canonical system of equations, allowing for real-time targeting and re-targeting procedures on-board. These procedures increase landing accuracy and enable entirely autonomous lunar descent guidance performance.

The contents of this book will interest scientists, engineers, and students working with autonomous GNC systems, the design and development of future landing missions, and the problems of planetary powered descent and precision landing, including lunar descent guidance. The book will benefit readers working in optimal control, applied mathematics, and mechanics in the development and utility of mission technologies. Readers are expected to have a familiarity with differential equations, the elements of motion in a central Newtonian gravitational field, and programming skills. The authors hope this book will also be helpful to those interested in the Apollo missions' heritage, the performance of Apollo guidance computers, and historical documents describing human landing missions to the Moon.

Honolulu, Hawai'i, USA Dilmurat M. Azimov
College Station, Texas, USA Robert H. Bishop
April 2025

Acknowledgements

We express our great appreciation to Mr. Alan Klumpp, the main author of lunar descent guidance algorithms and program and Mr. Don Eyles, a renowned engineer of the former MIT Instrumentation Laboratory (now known as the Charles Stark Draper Laboratory) and the hero of Apollo 14 mission, for their very helpful and positive discussions of the results of this monograph in 2010–2011.

As the first author, I note that the research which has been in progress since early 2006 and resulted in this monograph could not have been possible without a constant support of my family. Over these years, Mamura, Komola, Feruza, Dilshod and Rano Azimovs—my dear children, have been and will always be my endless source of courage, inspiration and motivations. Their help in typesetting of equations, proof readings, feedback, creating the figures and editorial suggestions are invaluable. With a great sense of gratitude and absolute sincerity I acknowledge a tremendous help and support of Mukhayyo Azimova. Her understanding and kindness have always been a critical factor in working with this monograph. I'm also grateful to my students, whose questions and comments on engineering mathematics of planetary landing maneuvers have undoubtedly polished the contents of this monograph. It is impossible not to mention the support that I received from my colleagues and friends in the course of work with this monograph. Especially, the countless hours of discussions and conversations with David Hull and Cesar Ocampo of the University of Texas at Austin, and Casey Monahan of the Texas Music Office always encouraged me towards the research on the utility of the Hamiltonian formalism in guidance and control and planetary landing problems. Finally, with a great pleasure I express my deep appreciation and gratefulness to Mukarram Azimova for her help in preparing the LaTex files and for her understanding and continuous support that she provided in the last few years of working with this monograph.

As the second author, I express my sincere gratitude to the students and colleagues who have contributed to my understanding of space trajectories over the years. During my tenure as an engineering staff member at the Charles Stark Draper Laboratory in the 1980s, I had the privilege of working with numerous Apollo-era engineers who mentored me and generously shared their expertise in space navigation and rendezvous with me, a newly minted engineer. The process of deciphering the Apollo entry guidance targeting and guidance from various fragmented documents was a delightful experience of discovery. The engineers of that era were exceptional educators. Additionally, I am thankful to Dilmurat Azimov (the first author of this manuscript) for the years spent collaborating on whiteboards and thoroughly examining the mathematics involved in searching for analytical solutions. My passion for the mathematics underlying our work in space trajectories, entry guidance, and targeting has been greatly enriched by the opportunity to dedicate countless hours to these topics.

Honolulu, Hawai'i, USA Dilmurat M. Azimov
College Station, Texas, USA Robert H. Bishop
April 2025

Contents

Part I Reconstruction of Apollo Ground-based Targeting and On-board Guidance for Powered Descent

1 Introduction 3
1.1 Returning to the Moon 3
1.2 Advancing the Guidance and Control Technology 4
1.3 Remarks on Reconstruction and Modification of Apollo Guidance 4
References 5

2 Apollo Descent and Landing Maneuver 7
References 9

3 Apollo Lunar Descent and Landing Guidance 11
3.1 Guidance: How a Polynomial Contributed to Land People on the Moon 11
3.2 Lunar-Landing Guidance Process 12
3.2.1 Process Overview 12
3.2.2 Nominal Lunar-Descent Trajectory 13
3.3 Guidance Cycle and Routines 16
References 17

4 Mathematical Preliminaries 19
4.1 Coordinate Systems 19
4.2 Transformation Matrices 22
4.2.1 Transformation Matrix from One System to Another 22
4.2.2 Determination of $\boldsymbol{\omega}$ and $\mathbf{R}_{LS}$ in Local and Platform Coordinate Systems 23
4.2.3 Transformation Matrix from Inertial to Local Coordinate System 23
4.2.4 Transformation Matrix from Inertial to Platform Coordinate System 23

4.2.5 Transformation Matrix from Local to Platform Coordinate System ... 24
4.2.6 Transformation Matrix from Local to Guidance Coordinate System ... 25
4.2.7 Supplementary Expressions ... 25
References ... 26

5 Approach-Phase Ground-Based Targeting ... 27
5.1 Approach-Phase Targeting Equations ... 27
5.1.1 Preliminary Expressions for the State Vector ... 27
5.1.2 The Approach-Phase Target, Initial and Terminal States ... 30
5.2 Approach-Phase Targeting Constraints, Inputs and Outputs ... 30
5.2.1 Constraints ... 31
5.2.2 Inputs and Outputs ... 31
5.3 Approach-Phase Targeting Algorithm ... 32
References ... 34

6 Braking-Phase Targeting ... 35
6.1 Introductory Remarks ... 35
6.2 Time and Mass Estimations ... 35
6.3 Estimation of Terminal Jerk and Snap Vectors for First Iteration ... 37
6.4 Expressions for Jerk and Snap Vectors ... 38
6.5 Braking-Phase Targeting Constraints, Inputs and Outputs ... 38
6.5.1 Terminal State Constraints ... 39
6.5.2 Operational Constraints ... 39
6.5.3 Inputs and Outputs ... 39
6.6 Braking-Phase Targeting Algorithm ... 39
6.7 Alternative Targeting Procedure Using Backward Integration ... 51
References ... 56

7 P63 Ignition Algorithm ... 57
7.1 Preliminary Comments ... 57
7.2 Utility of the Ignition and Guidance Algorithms ... 57
7.3 Modified Ignition Algorithm, Inputs and Outputs ... 58
7.3.1 Ignition Algorithm Inputs and Outputs ... 58
7.3.2 P63 On-Board Ignition Algorithm ... 59
Reference ... 61

8 P63-64 Powered Descent Guidance ... 63
8.1 Comments ... 63
8.2 Guidance Equations ... 64
8.2.1 Commanded Thrust Acceleration ... 64
8.2.2 Computation of Target Referenced Time ... 66

8.2.3 Computation of TPG_Y 66
8.3 P63-64 Guidance Algorithm, Inputs and Outputs 67
8.3.1 Algorithm Inputs and Outputs 67
8.3.2 Guidance Algorithm 68
Reference 73

9 P66 Lunar-Landing Guidance 75
9.1 P66 Horizontal Guidance Formula 75
9.2 P66 Vertical Guidance Formula 76
9.3 Commanded Body Frame Matrix in Attitude Routine 76
9.4 Concluding Remarks 77
Reference 78

10 Simulations and Comparison Analyses 79
10.1 Approach to Simulations 79
10.2 Comparison with Apollo 11 Post-Flight Data 80
10.3 Comparison with Apollo 12 Post-Flight Data 86
10.4 Conclusions 96
References 97

Part II Integrated Apollo-Class Real-Time Targeting and Guidance for Powered Descent

11 Extension and Enhancement of Apollo Guidance 101
11.1 Introductory Remarks 101
11.2 Enriching Lunar-Descent Guidance and Targeting with New Capabilities 102
11.3 Apollo Descent and Landing Maneuver Trajectory 102
11.4 Previous Studies 103
11.5 Extension and Enhancement of Guidance and Targeting 104
References 105

12 Optimal Control Problem and Extremals 107
12.1 Problem Statement 107
12.2 Extremals of the Problem 109
References 111

13 Extension of Braking-Phase Targeting 113
13.1 Braking-Phase Quartic 113
13.2 Closed-Form Solutions for Braking Phase 114
13.3 Initial and Final Conditions for Braking Phase 120
13.4 Continuity of Position and Velocity Vectors at TCR Point 121

13.5 Conditions for Vertical Component of Velocity Vector 123
13.6 Conditions for Horizontal Component of Velocity Vector 125
13.7 Position and Velocity Vectors in Polar Coordinate Frame 126
13.8 Platform Frame for a Nominal Landing Site 127
13.9 Position and Velocity Vectors in Platform Frame 127
References 128

14 Simulations of Braking Phase Using Analytical Solutions 129
14.1 Description of Simulations 129
14.2 Profiles of Thrust, Velocity and Flight Duration 138
14.2.1 Simulation Inputs 138
14.3 Sensitivity of Descent Trajectory to Initial Conditions 142
14.4 Simulations of Powered Descent at Moon South Pole 146
14.5 Concluding Remarks 146
References 150

15 Real-Time Targeting and Re-Targeting 151
15.1 Concept of Real-Time Targeting and Re-Targeting 151
15.2 Computation of Targeting Parameters 154
15.3 Comparison Analysis 155
15.4 Simulations of Descent and Landing at Shackleton Crater 156
15.5 Constraints, Inputs and Outputs 157
15.6 Analysis of Simulation Results 159
15.6.1 Qualitative Analysis 159
15.6.2 Numerical Results 160
15.7 Trajectory Parameters 160
15.8 Re-Targeted Trajectories 166
15.9 Concluding Remarks 167
References 168

16 Landing Accuracy 169
Reference 171

17 Conclusions 173
17.1 Reconstruction of Apollo Targeting and Guidance 173
17.2 Extension and Enhancement of Apollo-Class Targeting and Guidance 174

Nomenclature

$\mathbf{A}_{CP}$	Commanded (C) acceleration in platform (P) system
A_{TRIM}	Acceleration during trim phase, $[m/s^2]$
API	Approach-phase initial point
APT	Approach-phase target point
BRF	Braking-phase terminal point
BRI	Braking-phase initial point
BRT	Braking-phase target point
CBP	Commanded body frame matrix expressed in platform coordinates
Δt	Time interval between two consecutive cycles, integration time step $[s]$
$\mathbf{G}_{MP}$	Moon gravity vector in platform system
G_M	Magnitude of $\mathbf{G}_{MP}$, $[m/s^2]$
$\mathbf{J}_{BRTG}$	Jerk vector of braking phase target (T) point of in guidance (G) system
F_{BRF}	Thrust during braking phase throttle recovery, $[lbs]$
F_{THR}	Thrust during braking phase until throttle recovery, $[lbs]$
F_{max}	DPS maximum thrust, maximum allowable thrust $[lbs]$
$\dot{M}_{BRF}$	Mass-flow rate during throttle recovery period on braking phase, $[kg/s]$
$\dot{M}_{THR}$	Mass-flow rate on constant thrust arc of braking phase, $[kg/s]$
M_{TRIMF}	Final mass at terminal point of trim phase, $[kg]$
M_{BRF}	Final mass at terminal point of braking phase, $[kg/s]$
R_e, R_p	Moon mean equatorial and mean polar radii, $[m]$
R_M	Moon mean radius, $[m]$
$\mathbf{R}$, $\mathbf{V}$, $\mathbf{A}$, $\mathbf{J}$, $\mathbf{S}$	Position, velocity, acceleration, jerk and snap vectors respectively
$\mathbf{R}_I$	Position vector in inertial system (I)
$\mathbf{R}_{LSP}$	Landing site vector in platform system (P)

$\mathbf{R}_{APIG}$	Position vector of approach phase initial point in guidance system (G)
R_{XAPIG}	X-component of $\mathbf{R}_{APIG}$, $[m]$
$\mathbf{R}_{TG}$	Position of target (T) point in guidance system (G)
$\mathbf{R}_{LSL}$	Landing site vector, $\mathbf{R}_{LS}$, defined in local (L) system
STM	State transition matrix
$\mathbf{T}_A^B$	Matrix of transformation from A-frame to B-frame
$\mathbf{T}_P^L$	Matrix of transformation from platform to local Moon-Centered Moon-Fixed (MCMF) coordinate system
$\mathbf{T}_I^L$	Matrix of transformation from Moon-Centered inertial (MCI) to local MCMF coordinate system
$TAPF$	Target referenced time at approach-phase terminal point, $[s]$
$TAPI$	Target referenced time at approach-phase initial point, $[s]$
$TAPM$	Target referenced time at approach-phase middle point, $[s]$
$TAPT$	Target referenced time at approach-phase target point, $[s]$
$TBRI$	Target referenced time at braking-phase initial point (first P63 guidance pass), $[s]$
$TBRF$	Target referenced time at braking-phase terminal point, $[s]$
$TBRT$	Target referenced time at braking-phase target point, $[s]$
THR	Target referenced time at throttle control recovery point, $[s]$
$TRIM$	Duration of trim phase, $[s]$
$TRMI$	Time left until t_{LAND} at initial point of terminal descent phase, $[s]$
TTT	Target referenced time, $[s]$
$\mathbf{U}_{FCP}$	Unit thrust (F) vector, commanded (C) in platform (P) system
V_E	DPS exhaust velocity, $[m/s]$
$\mathbf{X}$	State vector
t_{CLOCK}	Clock-time, $[s]$
t_{BRI}	Initial clock-time of braking phase, $[s]$
t_{BRF}	Terminal clock-time of braking phase, $[s]$
t_I	Estimated clock-time of the first P63 guidance pass, $[s]$, initial clock-time of braking phase (full throttle)
t_{IG}	Initial clock-time of trim phase (ignition), $[s]$
t_{LAND}	Initial estimate of nominal landing time, $[s]$
t_{THR}	Estimated clock-time of initial point of throttle recovery, $[s]$
t_x	Ground elapsed time at the apolune of entry orbit (measured relative to lift-off time), $[s]$
t_0	Initial time of maneuver, $[s]$
$\mathbf{u}_{xip}$, $\mathbf{u}_{yip}$, $\mathbf{u}_{zip}$	Unit vectors used in computing $\mathbf{T}_I^P$
a_{trans}	Semi-major axis of entry orbit, $[m]$
e_{trans}	Eccentricity of entry orbit
p_{trans}	Semi-parameter of entry transfer orbit, $[m]$
ΔM_{TRIM}	Amount of fuel spent during trim phase, $[kg]$
Δt_{BA}	Duration of ballistic phase, $[s]$

Δt_{LTDI}	Time between the terminal descent phase initial point and t_{LAND}, $[s]$
Δt_{THR}	Estimated period of time between initial and end points of throttle recovery, $[s]$
Δt_{LTHR}	Time between t_{THR} and t_{LAND}, $[s]$
θ_{PBRF}	Angle between vertical X-axis of the platform system and thrust direction, $[deg]$
θ_{pbf}	Angle between vertical X axis of the platform system and thrust direction, $[deg]$
θ_s	Central angle between the inertial position vector and the landing site vector at current time, $[deg]$
ω_{MP}	Moon angular rotation vector in platform system
μ_M	Moon gravitational parameter, $[m^3/s^2]$
τ	Transport delay, effective P66 horizontal time constant, $[s]$
$\varphi_l,\ \theta_l$	Selenocentric latitude and longitude of landing site, $[deg]$
$\mathbf{g}_M$	Moon's gravity vector
F_{bf}	Thrust at final (f) point of braking (b) phase, $[N]$
F_{cf}	Thrust at final point of constant-thrust arc, $[N]$
$e_r,\ e_\theta$	Thrust direction cosines
g_0	Gravitational acceleration at sea level on Earth, $[m/s^2]$
h_p, h_a	Perilune and apolune of entry orbit, $[m]$
I_{sp}	Specific impulse, $[s]$
$m,\ m_0,\ m_1$	Mass, and initial and final masses, $[kg]$
m_{bf}	Mass at final point of braking phase, $[kg/s]$
r, v	Magnitudes of position and velocity vectors, $[m]$
r_p	Moon's mean polar radius, $[m]$
$\mathbf{r}_{agi}$	Position vector at initial (i) point of approach (a) phase in guidance (g) system
$\mathbf{r}_{big}$	Position vector at initial (i) point of braking (b) phase in guidance system
$\mathbf{r}_{bfg}$	Position vector at final (f) point of braking phase in guidance system
$\mathbf{r}_{btg}$	Position vector at target (t) point of braking phase in guidance system
$\mathbf{r}_{bxg}$	Position vector at initial point of braking-phase quartic in guidance system
$\mathbf{r}_p$	Position vector in platform (p) system
$\mathbf{r}_{lsl}^0$	Nominal landing site (ls) vector in local (l) system
$\mathbf{r}_{lsp}^0$	Nominal landing site vector in platform (p) system
$\mathbf{r}_{lsl}$	Current landing site vector in platform (p) system
$\mathbf{r}, \mathbf{v}, \mathbf{a}, \mathbf{j}, \mathbf{s}$	Position, velocity, acceleration, jerk and snap vectors respectively
r_{xaig}, r_{zaig}	X- and Z-components of $\mathbf{r}_{aig}$, $[m]$
r_{xamg}	X-component of position vector at middle (m) point of approach phase in guidance system, $[m]$

r_{bx}	Position vector magnitude at final point of constant-thrust arc, $[m]$
r_{bf}	Position vector magnitude at final point of braking phase, $[m]$
T	Target referenced time, $[s]$
TAM	Target referenced time at middle (M) point of approach (A) phase, $[s]$
TBI	Target referenced time at initial (I) point of braking (B) phase, $[s]$
TCR	Target referenced time at throttle control recovery (CR) point, $[s]$
$\mathbf{u}_{fcp}$	Unit vector of commanded (c) thrust (f) in platform (p) system
p_1	Coefficient of thrust percentage on constant-thrust arc
$\mathbf{x}$	State vector
$t,\ t_0,\ t_1$	Time, initial and final times of motion, $[s]$
t_f^0	Initial estimate of nominal landing time, $[s]$
t_x	Time at apolune of entry orbit, $[s]$
$u,\ w$	Radial and transversal components of velocity vector, $[m/s]$
$\mathbf{u}_{fcp}$	Unit vector of commanded (c) thrust (f) in platform (p) system at t_0
u_{fp}	Unit vector of thrust (f) in platform (p) system at t_0
Δt_{tdf}	Time of flight on terminal descent phase, $[s]$
Δt_e	Time of flight from apolune to perilune of entry orbit, $[s]$
Δt_{cf}	Time of flight on braking- phase quartic, $[s]$
β	Mass-flow rate, $[kg/s]$
$\lambda_u,\ \lambda_w,\ \lambda_r,\ \lambda_\theta,\ \lambda_m$	Lagrange multipliers for $u,\ w,\ r,\ \theta$ and m respectively, $\left[m/s, m^2/\left(s^3\right), m^2/\left(kgs^2\right)\right]$
μ	Moon's gravitational parameter, $[m^3/s^2]$
θ	Polar angle, $[rad]$

Part I
Reconstruction of Apollo Ground-based Targeting and On-board Guidance for Powered Descent

Chapter 1
Introduction

1.1 Returning to the Moon

Recently adopted "Space Policy Directive 1" emphasized on the development of innovative and sustainable program of space exploration. In particular, it stated the United States's willingness to return to the Moon followed by the missions to Mars and other destinations: "...Lead an innovative and sustainable program of exploration with commercial and international partners to enable human expansion across the solar system and to bring back to Earth new knowledge and opportunities. ......Beginning with missions beyond low-Earth orbit, the United States will lead the return of humans to the Moon for long-term exploration and utilization, followed by human missions to Mars and other destinations" [1]. These plans are illustrated through the development of the Gateway Program, which intends to provide the following capabilities: –50 kw -class power and propulsion element; supports exploration, science, and commercial activities in cislunar space and beyond; promotes international and US commercial development of elements and systems; provides options to transfer between cislunar orbits; and logistics and external robotic arm for berthing, science, exterior payloads, and inspections (see Fig. 1.1). These foundational gateway capabilities can support multiple US and international partner objectives in cislunar space and beyond [1]. The long-term goals of future Moon and Mars landing missions with advanced space technologies and past successful human and robotic landings suggest that understanding and taking full advantage of the heritage of the Apollo missions and landing maneuvers, including the knowledge and experience gained from Apollo lunar descent guidance concepts and related programs would be helpful [2, 3].

© The Author(s), under exclusive license to Springer Nature Switzerland AG 2026

D. M. Azimov and R. H. Bishop, *Apollo-Class Targeting and Guidance for Lunar Powered Descent and Precision Landing*, Springer Aerospace Technology,
https://doi.org/10.1007/978-3-031-91088-3_1

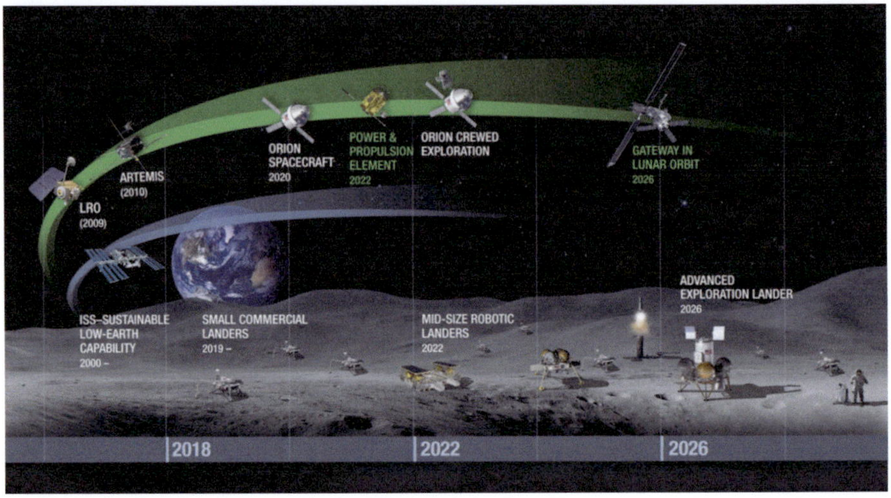

Fig. 1.1 Gateway program vision and illustration

1.2 Advancing the Guidance and Control Technology

This monograph presents the studies devoted to the reconstruction of the Apollo targeting and guidance algorithms executed to perform a powered descent and landing maneuver. The studies also aim to develop an integrated Apollo-class real-time targeting and guidance for potential applications to future autonomous and precision landing missions. Analysis of open literature and related technical reports show however, that the Apollo guidance concepts and their advantages have not been fully understood and utilized in the development of future landing guidance systems. The absence of a complete description of ground-based targeting, on-board ignition and guidance algorithms with necessary details of their integration and execution to provide safe landing adds more difficulty in advancing the autonomous guidance technologies for the future missions.

1.3 Remarks on Reconstruction and Modification of Apollo Guidance

The reconstruction of the original Apollo guidance program and advancing the guidance technology to obtain the performance parameters similar to the Apollo guidance performances would represent initial steps in the development of potential applications of this program to future missions. The purpose of this monograph is, in particular, to reconstruct the Apollo targeting algorithms utilizing the original guidance concepts along with details of their connection to ignition and guidance algorithms. These targeting algorithms are considered together with related functional procedures, terminal constraints, inputs and outputs, and are based on the original concepts, general definitions, conventions and the flow-chart diagrams employed in Apollo missions [4, 5]. The reconstructed algorithms are validated by comparing the simulation results with the post-flight trajectory and guidance data of the Apollo 11 and 12 missions. As an extension of the Apollo near-equatorial landing sites, the utility of the reconstructed algorithms have also been enhanced by successful application to a Moon South Pole mission with a landing site at Shackleton Crater [6].

References

1. Crystal Rosso (2018) NASA's gateway. MUREP other opportunities (MOO), p 16
2. Klumpp AR (1971) Apollo lunar-descent guidance, Apollo guidance navigation control, R-695. Massachusetts Institute of Technology, MIT Instrumentation Laboratory Report, Cambridge, MA
3. Apollo experience report—Mission planning for lunar module descent and ascent. NASA TN D-6846 (1972)
4. Guidance system operations plan for manned LM earth orbital and lunar missions using program luminary 1C (LM131 Rev. 1), section 5. Guidance equations. (Rev. 8). Apollo guidance navigation control, R-567. MIT, Cambridge, MA (1970)
5. Project Apollo coordinate system standards (1965) Apollo program, SE 008-001-1, NASA office of manned space flight. http://ntrs.nasa.gov/archive/nasa/casi.ntrs.nasa.gov/19700076120_1970076120.pdf
6. Azimov DM, Bishop RH (2008) Reconstruction of Apollo targeting and guidance for powered lunar descent. In: New trends in astrodynamics and applications V, Milan, Italy

Chapter 2
Apollo Descent and Landing Maneuver

A typical Apollo lunar descent and landing maneuver started with a transfer from an initial circular lunar orbit at an altitude of 100 km above the Moon"s surface to an elliptical entry orbit at its apolune by means of a finite thrust as shown in Fig. 2.1 of the Apollo 12$'s$ maneuver design [1]. The insertion into an approximately 15.3 km $\times$ 100 km elliptical transfer orbit was a common maneuver in the sequence of maneuver operations in all Apollo missions (see Fig. 2.2). The Apollo descent and landing maneuver includes ballistic, ullage, trim, braking, approach and terminal descent phases. The maneuver trajectory is nearly planar, and a significant part of the powered descent trajectory is approximated by two quartic polynomials, the approach-phase (AP) quartic and braking-phase (BP) quartic polynomials as shown in Fig. 3.2 [2]. The AP-quartic is defined in a closed form, while the determination of the BP-quartic requires iterations in order to satisfy terminal constraints. No closed-form solution was utilized on the significant part of Apollo BP from the starting point until the throttle control recovery (TCR) point. The use of a quartic for a nominal trajectory approximation is reasoned by the fact that its formulation is of a minimum order to satisfy initial and terminal constraints between the phases [2, 3]. The transfer from the circular orbit into the elliptical entry orbit is followed by ballistic flight until the lander enters the neighborhood of perilune of the entry orbit where the powered descent maneuver is initiated. Before reaching the perilune, the lander initiates ullage for 7.5 s followed by a trim (TR) phase with minimum thrust, typically, of $10\% F_{max}$ of the descent propulsion system (DPS) (see Fig. 3.2 of Chap. 3). After approximately 26 s of motion with the minimum thrust, near or at the perilune of the entry orbit, the lander starts its DPS initiates braking phase with constant thrust (CT), typically with $93-94\% F_{max}$. At TCR, when the thrust level is decreased to about $57\% F_{max}$, this phase is changed to throttle recovery, which lasts for 120 s. The trajectory during throttle recovery is approximated by the BP-quartic polynomial. The approach phase,

© The Author(s), under exclusive license to Springer Nature Switzerland AG 2026

D. M. Azimov and R. H. Bishop, *Apollo-Class Targeting and Guidance for Lunar Powered Descent and Precision Landing*, Springer Aerospace Technology,
https://doi.org/10.1007/978-3-031-91088-3_2

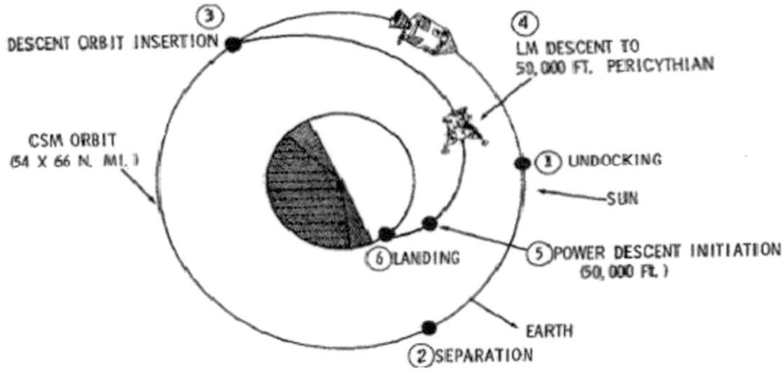

Fig. 2.1 Lunar descent events (Apollo 12) [1]

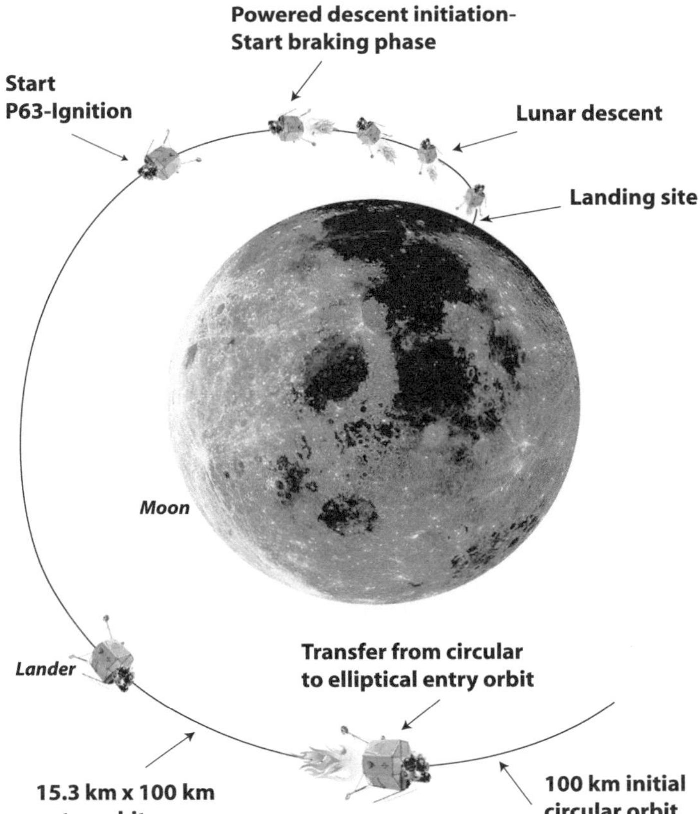

Fig. 2.2 Apollo-driven powered descent and landing scenario: The lander is transferred from a 100 km. initial circular parking orbit into a 15.3 km × 100 km entry orbit, and the powered descent initiation takes place near the perilune of the entry orbit [1]

described by the AP-quartic, lasts up to 160 s and ends at the initial point of the terminal descent (TD) phase at about 30 m altitude and 11 m ground range to landing site and at about 30 s to touchdown. The TD phase is nearly vertical, the horizontal velocity is nulled, and further targeting procedure is not performed during this phase.

References

1. Apollo 12 Mission Report (1970) Trajectory reconstruction and post-flight analysis, vol. 1. NASA-TM-X-70272, Houston, Texas, p 113
2. Klumpp AR (1971) Apollo lunar-descent guidance, Apollo guidance navigation control, R-695. Massachusetts Institute of Technology, MIT Instrumentation Laboratory Report, Cambridge, MA
3. Yang TL (1971) A targeting scheme for fuel optimal rocket trajectories with applications to the LM descent braking phase. Technical memorandum TM—71-2014-1, Bellcomm

Chapter 3
Apollo Lunar Descent and Landing Guidance

3.1 Guidance: How a Polynomial Contributed to Land People on the Moon

As a process, guidance can be defined as follows. Guidance is a nonlinear control process in which a vehicle is guided from its initial or current state to a specified target state. The guidance process is based on a guidance law and it is implemented by a guidance system which includes spacecraft's on-board computer, actuators and measurement units that provide measured quantities. During this process an on-board computer keeps the spacecraft close to reference (or nominal trajectory) through processing the guidance measurements. In this section we describe the historical background of guidance, that is how the theory of guidance was developed, and how a fourth-order polynomial contributed to land people on the Moon.

The guidance theory in the context of an optimization problem was reported by Kelley in 1961 [1]. In 1966, this theory was further examined by Hu and Thompson to develop one of the first analytical methods describing guidance functions in the form of polynomials [2]. A general explicit guidance methodology known as the Explicit (E)-guidance was proposed by George Cherry as a different approach to the guidance problem [3]. Later on, Moore et al. reported an implicit guidance equations [4]. Alan Klumpp simplified Cherry's guidance equations for Apollo Lunar Module (LM) guidance computer (LGC) coding and generalized it to the n-th order. Based on Cherry's methodology, Klumpp formulated the Lunar-descent guidance in terms of polynomials which was then implemented during the Apollo 11 and subsequent Apollo missions [5–8]. Defining the LM's position vector as a fourth-order polynomial to describe two critically important phases of the powered descent trajectory allowed for the derivation of the predictive guidance equation in simple and easily understandable terms. This guidance equation was then utilized in Programs 63 and 64 (P63 and P64). The Program 66 (P66) with vertical and horizontal channel design for the last phase of the trajectory was developed by Widnall and by Schulenberg and Suddath [5, 9]. Note that the core algorithms of the P63 and P64

© The Author(s), under exclusive license to Springer Nature Switzerland AG 2026

D. M. Azimov and R. H. Bishop, *Apollo-Class Targeting and Guidance for Lunar Powered Descent and Precision Landing*, Springer Aerospace Technology,
https://doi.org/10.1007/978-3-031-91088-3_3

guidance programs were designed using two fourth-order polynomials, and it is quite remarkable that these programs along with P66 landed astronauts on the Moon. The Lunar Digital Autopilot was designed by Widnall and Keene, and the entire Apollo LM GC software for the final lunar descent was developed by Eyles [9, 10].

The corresponding guidance algorithms were described in the Colossus 1 through Colossus 3 programs and Luminary 1A through 1E programs [11, 12]. The Colossus programs were installed in the Command Module (CM) and were run on the Apollo Guidance Computer (AGC). Sundisk was a pre-Colossus version of Command Module AGC's software. There were minor versions such as Colossus 1A, Colossus 2C, etc. Colossus 2 was also referred to as Comanche (used in Apollo 11 and Apollo 13). The Luminary 1A-1E programs were implemented on the LM GC [5, 8, 13]. The main guidance programs were P63 for ignition, P63-64 for guidance on braking and approach phases, P66 for horizontal guidance and P66 for vertical guidance on terminal descent phase, and also throttle and attitude control programs. Navigation solutions provided a current state vector's estimate by utilizing data from the inertial measurement unit (IMU) and a landing radar [14]. Although in most cases the existing literature and reports include some detailed descriptions of the programs, the questions on their connection with targeting and other algorithms, and the basic step-by-step guidelines of their execution have not been reported or thoroughly studied with the purpose of analyzing potential applications to future mission technologies, such as ALHAT (Autonomous Landing and Hazard Detection Technology) or the autonomous guidance and control system design for safe and precision landing on Mars and other celestial bodies [15, 16].

3.2 Lunar-Landing Guidance Process

3.2.1 Process Overview

Following Klumpp's document on Apollo lunar-descent guidance and the post-mission documents, one can describe the landing guidance performance and its elements as follows [5, 8]. Lunar-landing guidance is divided into two distinct parts: pre-ignition phase (Ignition algorithm) and guidance cycle starting before ullage initiation and ending after touchdown. The Lunar-descent guidance is performed by the on-board LM guidance computer (LGC) by controlling attitude and thrust (see Fig. 3.1). The primary guidance mode is automatic. The guidance process begins at about 15 km altitude in an elliptical coasting lunar orbit and end with landing the Lunar Module (LM) on the lunar surface.

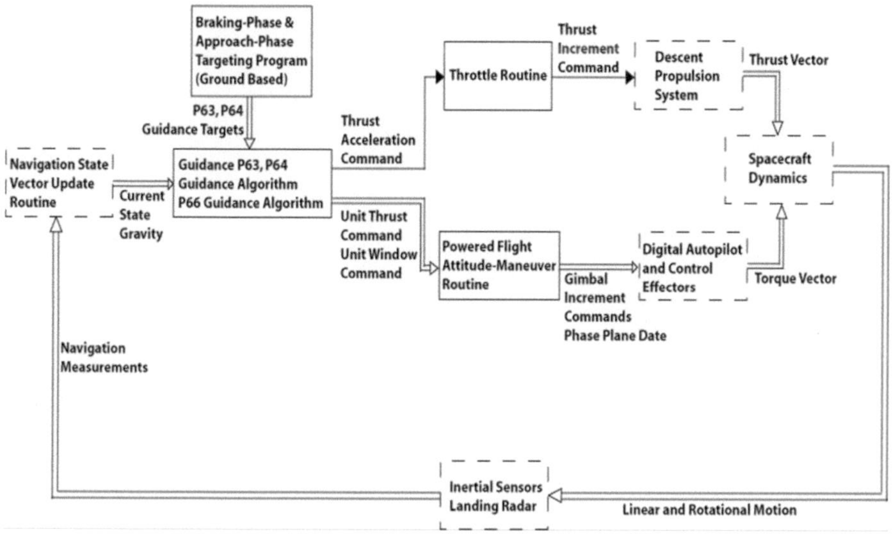

Fig. 3.1 Original Apollo GNC configuration diagram [5]

3.2.2 Nominal Lunar-Descent Trajectory

A nominal lunar-descent trajectory is planar trajectory and consists of the following phases (see Fig. 3.2):

- Pre-ignition
- Ullage and Trim
- Braking phase
- Approach phase
- Terminal descent-phase.

Pre-Ignition Phase. The pre-ignition phase starts in the elliptical descent orbit. This phase determines the ignition time based on the LM state in guidance coordinates. By the first guidance cycle of the braking phase, when the execution of P63 is initiated shortly before the initiation of ullage, the unit vectors defining the guidance frame are determined, and the thrust acceleration and window-pointing vectors are commanded. The LM state vector can be maintained using accelerometer measurements (average-G equations). When the determination of the guidance frame and these vectors is complete, the LM state vector is extrapolated forward to the time at which the guidance process is initiated. During this phase the LM travels in a ballistic trajectory with no thrust. The attitude maneuver routine orients the LM for the succeeding ullage and trim phase such that the X -axis lies in the direction of the first thrust command predicted by the pre-ignition computations.

Ullage and trim. The ullage and trim phase takes place as a continuation of P63 ignition program execution. The LGC master ignition routine starts Reaction Control

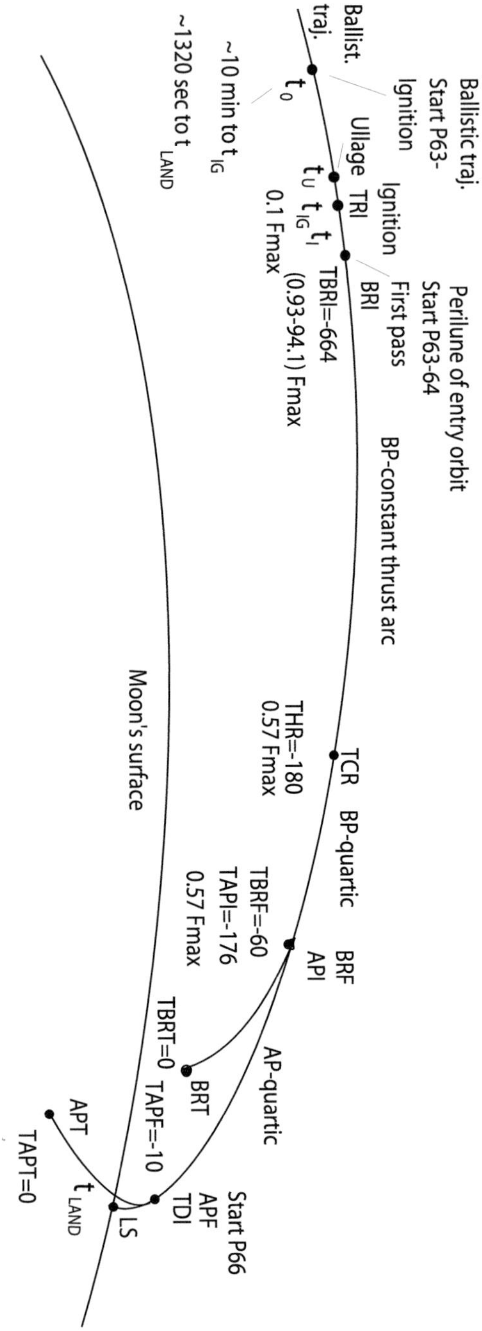

Fig. 3.2 Original design of Apollo nominal descent and landing trajectory with guidance programs [5]

System (RCS) ullage and shortly thereafter ignites the engine at minimum throttle setting, 10% of maximum thrust for a period of time sufficient to trim the thrust vector to point through the center of mass. The engine maintains a constant minimum throttle setting for about 26 s and the LM maintains a constant attitude throughout this phase.

Braking phase. As shown in Fig. 3.2, the braking phase is initiated by keyboard entry about 10 min before the nominal ignition time starting with the first guidance pass and controlled by P63 until the BP-terminal point: 22 min and 45° central angle to touchdown. The throttle is commanded to maximum at the precise instant computed by the pre-ignition computations for starting the braking phase. To utilize propellant efficiently, as it was later described Yang using optimal control theory, it is necessary to operate the Descent Propulsion System (DPS) at maximum thrust as long as possible [17]. The throttle command logic holds the throttle at maximum until the command thrust acceleration is less than a pre-selected value. By properly selecting the BP targets, the throttling -down is made at a specified time before the end of the phase on a nominal trajectory. The state vector updates are based on inertial measurement unit (IMU) data until estimated altitude is below a pre-selected value. After this time, the landing radar altitude measurements are used to update the LM's state vector. The velocity measurements are used when estimated speed is below pre-selected value. The landing radar provides altitude data below 10 km altitude, and velocity data below 610 m/s. At typically 492 km slant range from the LM to the landing site, P63 ignites the DPS. The P63 transfers LM to the terminal state within typically 514 s. and the LM's trajectory is planar and nearly optimal in terms of fuel expenditure. After about 11 min 31 s and having travelled 492 km from the ignition point, the LM will land at a location with 16.2° central angle. The phase ends at 2231 m altitude, 7471 m ground range and 177 s before touchdown. See Chap. 4 for more details regarding the nominal trajectory, the coordinate frames and the central angle.

Approach phase. The initial state and the targets for this phase are chosen so that the LM's X-axis (the thrust vector) is elevated sufficiently high above the local horizontal to permit landing site visibility. The DPS is operated in the continuously thottleable region throughout the approach phase. The new landing site location is computed in each guidance cycle. The approach phase is controlled by P64 and it begins with the LM at about 2.2 km altitude still moving about 129 m/s forward velocity and 44 m/s downward velocity. In typically 146 s, P64 transfers the LM to a point almost directly above the landing site providing continuous visibility of the lunar surface. The terminal point is at about 30 m altitude, 11 m ground range and 31 s to touchdown.

Terminal-Descent phase. The terminal -descent phase is controlled by P66 and it begins automatically at the terminal point of the approach phase. Note that, theoretically, it may also be initiated any time during the approach phase. The P66 consists of P66 Horizontal Guidance and P66 Vertical (ROD) Guidance Algorithms. The P66 Horizontal Guidance algorithm is processed every two seconds, based on assumption that the vertical component of thrust acceleration is equal to the gravitational acceleration on the Moon and it nulls LM's horizontal velocity with respect to the lunar surface and to produce a vertical approach to it. The ROD is processed every second, it controls the LM's vertical velocity only (altitude rate) to a reference value that is incremented or decremented by 0.3 m/s each time and does not control its attitude.

3.3 Guidance Cycle and Routines

In general, the guidance cycle produces (see Fig. 3.1) [5]:

- instant of time at which the state vector is estimated,
- the state vector,
- commanded thrust acceleration and window -pointing vectors, based on the state vector, and satisfying the constraints of the current phase,
- throttle commands, based on commanded thrust acceleration,
- digital autopilot (DAP) commands, based on the commanded thrust acceleration and window-pointing vectors.

Guidance transfers the LM from the current state to the terminus of the current phase. In addition to the current state estimate from Navigation, Guidance is based on ground-based pre-computed targets from the targeting program. The outputs of the guidance algorithm are a unit thrust command and a unit window command issued to Attitude maneuver routine, and a thrust acceleration command, issued to Throttle routine. Except for the P63, P64 and P66 programs, the guidance system includes three more basic routines: Powered-flight attitude maneuver routine, Throttle routine and the BP- targeting and AP-targeting programs.

Powered-flight attitude maneuver routine. This routine connects all programs to the DAP. It transfers the LM from any current attitude to any commanded attitude. The routine inputs are the unit thrust command and the unit window command. It estimates the unit thrust vector from accelerometer measurements, and issues incremental commands to the DAP. These commands cause the DAP to drive the estimated unit thrust vector into coincidence with the unit thrust command and the symmetry plane of the LM into coincidence with the unit window command. Typically, the unit window command is the line-of-site vector from the LM to the current landing site (see Fig. 3.3).

Throttle routine. This routine connects several guidance programs to the DPS. The DPS must be operated either at maximum thrust point or within a permitted thrust region (11%–65%). The region (65%–93%) is forbidden because in this region "the oxidizer flow and fuel flow make independent transitions from caviating to noncaviating regimes". The routine computes the current and commanded thrusts and issues thrust increment commands to the DPS [5].

BP-targeting and AP-targeting programs. The targeting program generates mission-dependent data for the P63 Ignition algorithm, P63 and P64 guidance algorithms. These algorithms require targets for P63 that provide an efficient transfer and

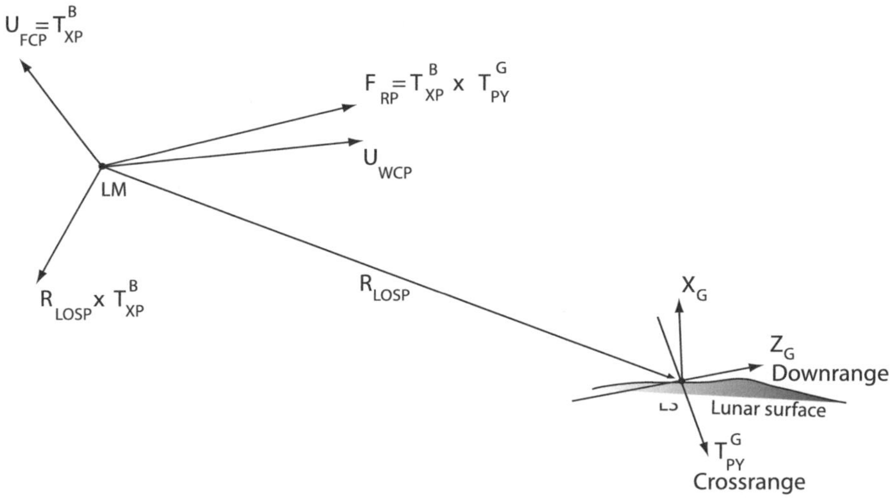

Fig. 3.3 Illustration of the commanded thrust and commanded window vectors

targets for P64 that provide a trajectory meeting several constraints on geometry, visibility and thrust. The task of the targeting program is to set up the P64 initial state and guidance targets such that suitable visibility and thrust profiles are realized implicitly. During the final portion of P63, and throughout P64, the guidance will generate a trajectory in which the LM's position vector is a quartic polynomial function of time. The targeting programs define each of the BP-quartic and AP-quartic polynomials and extract the guidance targets as the position vector and its derivatives at a target point, lying on the polynomial, beyond the phase terminus [5].

References

1. Kelley HJ (1962) Guidance theory and extremal fields. IEEE Trans Autom Control 5(2):75–82
2. Hu S, Thompson M (1966) A direct and analytical solution for space flight guidance functions. Northrop Space Laboratories, Huntsville, AL
3. Cherry GW (1964) E-guidance—A general explicit, optimizing guidance law for rocket-propelled spacecraft, R-456. Massachusetts Institute of Technology, MIT Instrumentation Laboratory Report, Cambridge, MA
4. Moore TE, McSwain GG, Montgomery JD (1969) Guidance laws for controlling off-nominal LM powered descent trajectories back to the nominal. Internal note MSC-EG-69-9, Project Apollo, NASA, Manned Spacecraft Center, Houston, Texas
5. Klumpp AR (1971) Apollo lunar-descent guidance, Apollo guidance navigation control, R-695. Massachusetts Institute of Technology, MIT Instrumentation Laboratory Report, Cambridge, MA
6. Apollo experience report—mission planning for lunar module descent and ascent, NASA TN D-6846 (1972)
7. Apollo mission 11 (1970) Trajectory reconstruction and post-flight analysis, vol 1. Project Apollo. TASK MSC/TRW A-50. NAS 9-8166, Houston, Texas, p 142

8. Apollo 12 mission report (1970) Trajectory reconstruction and post-flight analysis, vol 1. NASA-TM-X-70272, Houston, Texas, p 113
9. Widnall WS (2012) Lunar module digital autopilot. J Spacecr 8(1). https://doi.org/10.2514/3.30217
10. Eyles DE (1973) Apollo LM guidance computer software for the final lunar descent. Automatica 9(2):243–250. https://doi.org/10.1016/0005-1098(73)90078-2
11. Guidance system operations plan for manned LM earth orbital and lunar missions using program luminary 1C (LM131 Rev. 1), section 5. Guidance equations. (Rev. 8). Apollo guidance navigation control, R-567. MIT, Cambridge, MA (1970)
12. Guidance system operations plan for manned LM earth orbital and lunar missions using program colossus 3 (LM131 Rev. 1), section 5. Guidance equations. (Rev. 14). Apollo guidance navigation control, R-577. MIT, Cambridge, MA (1971)
13. Apollo 11 lunar landing mission. Press Kit. NASA, p 100 (1969)
14. Kriegsman BA (1966) Radar-updated inertial navigation of a continuously- powered space vehicle. In: IEEE aerospace systems conference, Seattle, Washington
15. Epp CD, Smith TB (2008) The autonomous precision landing and hazard detection and avoidance technology (ALHAT). In: Go for lunar landing: from terminal descent to touchdown conference, panel 4: GNC. http://ser.sese.asu.edu/GO/
16. Wong EC, Singh G, Masciarelli JP (2002) Autonomous guidance and control design for hazard avoidance and safe landing on mars. In: AIAA atmospheric flight mechanics conference, Monterey, CA
17. Yang TL (1971) A targeting scheme for fuel optimal rocket trajectories with applications to the LM descent braking phase. Technical memorandum TM—71-2014-1, Bellcomm

Chapter 4
Mathematical Preliminaries

4.1 Coordinate Systems

The five coordinate systems used in the studies are: the selenocentric inertial (Moon-centered), the selenocentric local (Moon-fixed), the selenocentric platform, the guidance and the body coordinate systems. Note that the Y-axis completes the right-handed triad in all these coordinate systems [1]. In this work, the subscripts or superscripts L, P, G, B, I will indicate the local, platform, guidance, body and inertial systems respectively. These systems are depicted in Figs. 4.1, 4.2, 4.3.

Selenocentric inertial coordinate system. The origin of the inertial coordinate system is located at the Moon center of mass (COM). The X_I-axis is aligned along the mean vernal equinox (intersection of the Earth mean equatorial plane and the Earth mean orbit, e.g., the ecliptic) at a fixed epoch (January 1, 2000). The Z_I-axis is defined along the Earth mean rotational axis.

Selenocentric local coordinate system. This system is an orthogonal, Moon-centered Moon-fixed (MCMF) system with the origin at the Moon COM (see Fig. 4.2). The X_L-axis is along the intersection of the meridian of 0° longitude and the Moon mean equatorial plane. This coordinate system is defined by the axis of rotation that lies on the north side of the invariable plane of the solar system and is specified by its right ascension and declination, represented by α_0 and β_0, respectively, as illustrated in Fig. 4.3. The prime meridian is specified by the twist angle, denoted by W, that is measured along the planet equator in an easternly direction as specified by the angle from the node Q to the point where the prime meridian crosses the equator. Therefore, the location of the north pole and the intersection of the prime meridian with the equator are determined by α_0, β_0, and W. The Z_L-axis is along the Moon rotational axis.

© The Author(s), under exclusive license to Springer Nature Switzerland AG 2026

D. M. Azimov and R. H. Bishop, *Apollo-Class Targeting and Guidance for Lunar Powered Descent and Precision Landing*, Springer Aerospace Technology,
https://doi.org/10.1007/978-3-031-91088-3_4

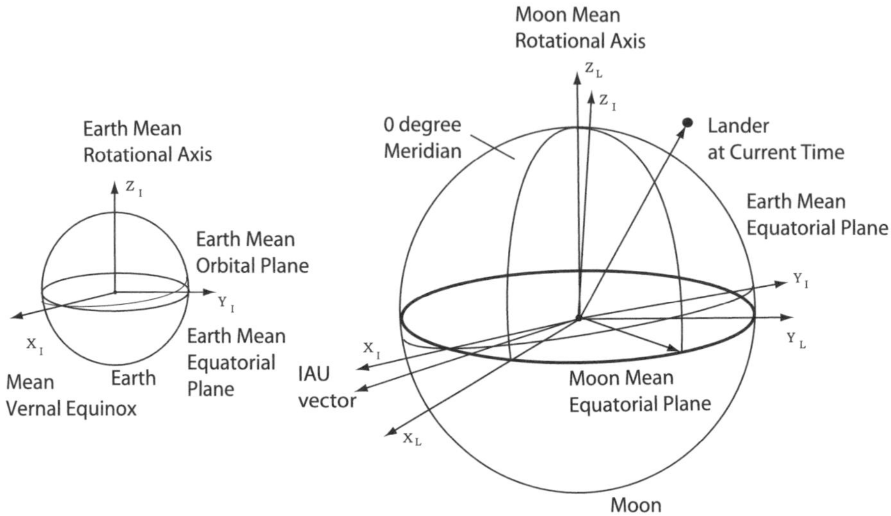

Fig. 4.1 Selenocentric inertial (Moon-centered) and selenocentric local (Moon-fixed) coordinate systems

Selenocentric platform coordinate system. This system is defined as an orthogonal non-rotating system with the origin at the Moon COM (see Fig. 4.2). The X_P-axis is aligned towards the nominal LS at the nominal landing time. The Z_P-axis lies in the orbital plane and is along the direction of the lander's motion. Note that the nominal landing trajectory lies in the $X_P Z_P$-plane. This plane contains the lander in its current position, the nominal LS at nominal landing time and the Moon COM. Therefore, the initial and entry orbits also lie in this plane.

Guidance coordinate system. This system is an orthogonal system whose origin coincides continuously with the current LS (see Fig. 4.2). The X_G-axis is vertical and points upward. The Z_G-axis lies in the instantaneous $X_G Z_G$-plane containing the lander in its current position, the current LS and the Moon COM, and is aligned along the direction of the lander's motion.

Body coordinate system. The coordinates of the body coordinate system can be considered as the coordinates of the lander. The origin of the body coordinate system is located 200 inches below the lander's ascent stage base. The X_B-axis is aligned along the current thrust vector. The Z_B-axis is parallel to the centerline vector of the exit hatch and is directed forward from the design eye. This system is not used in the targeting program, and its description is provided here for completeness.

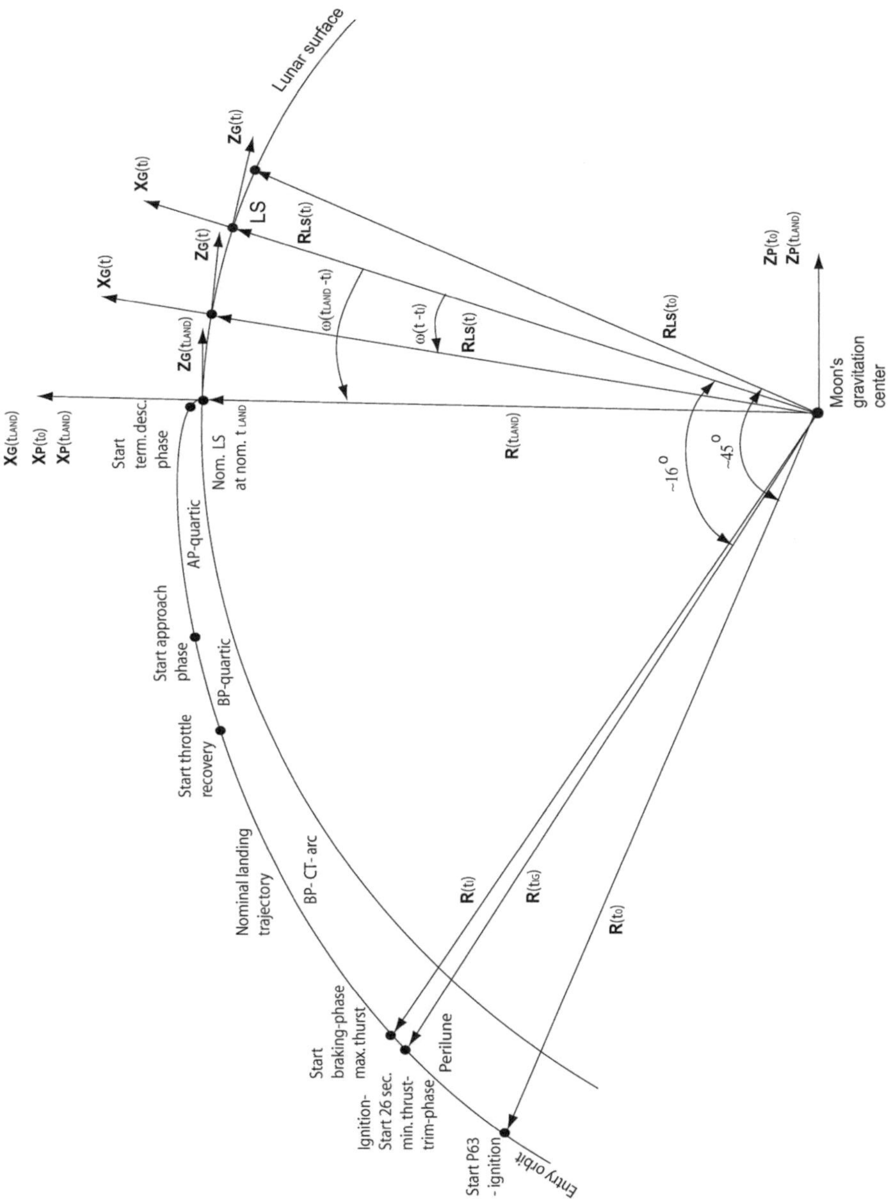

Fig. 4.2 Guidance and platform coordinate systems and nominal lunar-descent and landing trajectory

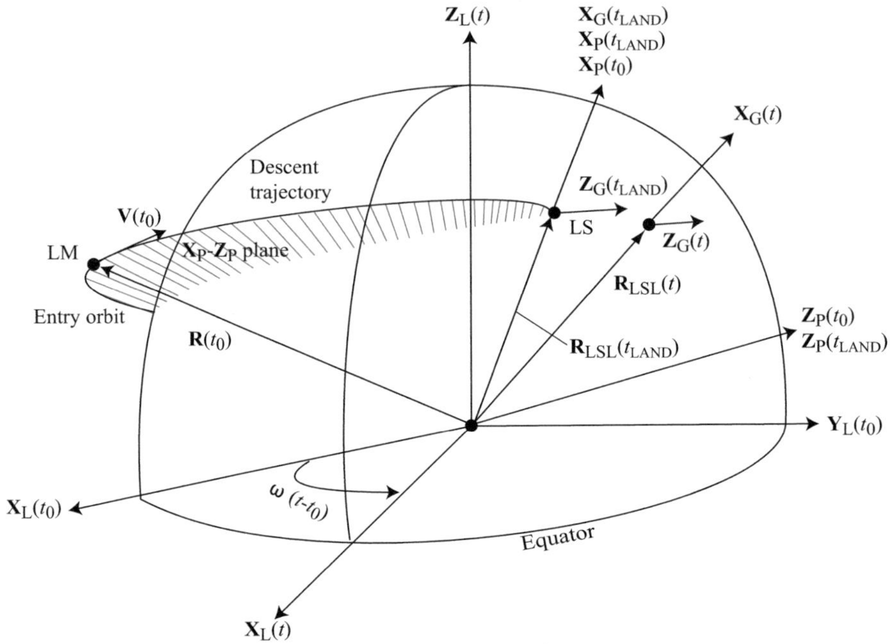

Fig. 4.3 Local, platform and guidance coordinate systems and nominal landing trajectory

4.2 Transformation Matrices

4.2.1 Transformation Matrix from One System to Another

It is known that if $\mathbf{a}$, $\mathbf{b}$ and $\mathbf{a}'$, $\mathbf{b}'$ are two unit vectors defined in two cartesian systems $Oxyz$ and $Ox'y'z'$ respectively, then one can construct a transformation matrix from one system to another [2]:

$$\mathbf{T}_{Oxyz}^{Ox'y'z'} = \left[\mathbf{X}^T,\ \mathbf{Y}^T,\ \mathbf{Z}^T\right]. \tag{4.1}$$

where

$$\begin{aligned}
\mathbf{X} &= u'_{xx}\mathbf{u}_x + u'_{yx}\mathbf{u}_y + u'_{zx}\mathbf{u}_z,\\
\mathbf{Y} &= u'_{xy}\mathbf{u}_x + u'_{yy}\mathbf{u}_y + u'_{zy}\mathbf{u}_z,\\
\mathbf{Z} &= u'_{xz}\mathbf{u}_x + u'_{yz}\mathbf{u}_y + u'_{zz}\mathbf{u}_z,
\end{aligned} \tag{4.2}$$

$$
\begin{aligned}
\mathbf{u}_x &= \mathbf{a}, \\
\mathbf{u}_y &= unit[\mathbf{a} \times \mathbf{b}], \\
\mathbf{u}_z &= \mathbf{u}_x \times \mathbf{u}_y, \\
\mathbf{u}'_x &= \mathbf{a}', \\
\mathbf{u}'_y &= unit[\mathbf{a}' \times \mathbf{b}'], \\
\mathbf{u}'_z &= \mathbf{u}'_x \times \mathbf{u}'_y.
\end{aligned} \tag{4.3}
$$

4.2.2 Determination of ω and $\mathbf{R}_{LS}$ in Local and Platform Coordinate Systems

The vectors $\boldsymbol{\omega}$ and $\mathbf{R}_{LS}$ can be determined in the local system as follows:

$$\boldsymbol{\omega}_L = [0\;0\;\omega]^T, \quad \mathbf{R}_{LSL} = [R_M \cos\varphi \cos\theta \quad R_M \cos\varphi \sin\theta \quad R_M \sin\varphi]^T. \tag{4.4}$$

In the platform coordinate system, these vectors can be computed as

$$\boldsymbol{\omega}_P = \mathbf{T}_L^P \boldsymbol{\omega}_L, \quad \mathbf{R}_{LSP} = \mathbf{T}_L^P \mathbf{R}_{LSL}. \tag{4.5}$$

4.2.3 Transformation Matrix from Inertial to Local Coordinate System

It will be assumed that $\mathbf{T}_I^L$ can be computed at any time during the maneuver. For its computation the reader is referred to the Apollo Project Standard Coordinate Systems with Planetary Inertial Orientation Subroutine, or to the report of the IAU/IAG Working Group which defines the Moon Orientation based on the Earth Mean Equator of J2000 and Moon Equator [1, 3].

4.2.4 Transformation Matrix from Inertial to Platform Coordinate System

Assume that the orientation of the plane in which the landing maneuver is to take place or the lander's angular momentum vector, $\mathbf{h}$ known at t_x (the instant of time when the lander is at apolune of entry orbit), can be given as

$$\mathbf{h}(t_x) = \mathbf{R}_I(t_x) \times \mathbf{V}_I(t_x).$$

This means that the plane of the landing maneuver is determined as perpendicular to the angular momentum vector, $\mathbf{h}$ at t_x and contains both $\mathbf{R}_I(t_x)$ and $\mathbf{V}_I(t_x)$. From the definition of the selenocentric platform coordinate system one can see that this plane is the $X_P Z_P$-plane of this coordinate system, which coincides with the entry orbit's plane and contains the following three points at any time throughout the maneuver starting from t_x: the lander's COM in its current position, the nominal LS at nominal landing time and the Moon COM (see Fig. 4.2).

Note. Another approach to determining the plane in which the landing maneuver would be going to take place would be an assumption that four orbital elements of the entry orbit, that is the semi-parameter, eccentricity, inclination and longitude of ascending node be known at t_x. Analysis show that the assumption that the lander's angular momentum vector is known at t_x is convenient in describing the maneuver plane in connection with the trajectory and guidance parameters. It is also assumed that the LS vector, $\mathbf{R}_{LSL}$ is known in the MSMF local coordinate system. Then the constant matrix $\mathbf{T}_I^P$ is given by

$$\mathbf{T}_I^P = \mathbf{T}_I^P(t_x) = \mathbf{T}_I^P(t_{LAND}) = \left[\mathbf{u}_{xip}^T,\ \mathbf{u}_{yip}^T, \mathbf{u}_{zip}^T\right], \tag{4.6}$$

where the unit vectors of the platform coordinate axis are

$$\begin{aligned}
\mathbf{u}_{xip} &= \frac{\mathbf{R}_{LSI}(t_{LAND})}{||\mathbf{R}_{LSI}(t_{LAND})||}, \\
\mathbf{u}_{zip} &= \frac{\mathbf{h}(t_x) \times \mathbf{u}_{xip}}{||\mathbf{h}(t_x) \times \mathbf{u}_{xip}||}, \\
\mathbf{u}_{yip} &= \mathbf{u}_{zip} \times \mathbf{u}_{xip}
\end{aligned} \tag{4.7}$$

and

$$\mathbf{R}_{LSI}(t_{LAND}) = \mathbf{T}_L^I(t_{LAND})\mathbf{R}_{LSL}.$$

Note that if $\mathbf{R}_I(t_x)$ and $\mathbf{V}_I(t_x)$ or the lander's transfer orbit parameters and its location is this orbit are known at t_x, then $\mathbf{T}_I^P$ can be computed on-board. However, if these position and velocity vectors or the information about the lander's transfer orbital plane are unknown, then it may necessary to compute $\mathbf{T}_I^P$ utilizing some other methods in order to run the targeting algorithm on the ground before the mission.

4.2.5 *Transformation Matrix from Local to Platform Coordinate System*

It can be shown that the matrix $\mathbf{T}_I^P$ may be computed in the form:

$$\mathbf{T}_I^P = \mathbf{T}_I^P(t_{LAND}) = \mathbf{T}_L^P(t_{LAND})\mathbf{T}_I^L(t_{LAND}). \tag{4.8}$$

Then one can compute $\mathbf{T}_L^P(t),\ \forall t > t_I$ and $\mathbf{T}_L^P(t_{LAND})$ as follows:

$$\mathbf{T}_L^P(t) = \mathbf{T}_I^P \left[\mathbf{T}_I^L(t)\right]^T \tag{4.9}$$

$$\mathbf{T}_L^P(t_{LAND}) = \mathbf{T}_I^P \left[\mathbf{T}_I^L(t_{LAND})\right]^T \tag{4.10}$$

4.2.6 Transformation Matrix from Local to Guidance Coordinate System

The matrix $\mathbf{T}_L^G$ is computed at the first guidance call at time t_I. At this time, for the first iteration in the guidance subroutine, it is assumed that $\mathbf{T}_P^G(t_I) = \mathbf{I}$. Then

$$\mathbf{T}_L^G(t) = \mathbf{T}_L^G\left(\theta_L, \varphi_L, \mathbf{R}(t), \mathbf{V}(t)\right) = \mathbf{T}_P^G(t)(\mathbf{T}_P^L(t))^T, \qquad (\forall t > t_I). \tag{4.11}$$

4.2.7 Supplementary Expressions

The matrix $\mathbf{T}_L^P$ can be computed according to Eq. (4.1):

$$\mathbf{T}_L^P(t_{LAND}) = \left[\mathbf{X}^T\ \mathbf{Y}^T\ \mathbf{Z}^T\right]^T. \tag{4.12}$$

Consider the angular momentum vector $(\mathbf{R}_I \times \mathbf{V}_I)$ used in $\mathbf{T}_I^P$:

$$\mathbf{T}_I^P = [a_{ij}] = \left(\mathbf{u}_x^T\ \mathbf{u}_y^T\ \mathbf{u}_z^T\right)^T, \qquad i, j = 1, 2, 3. \tag{4.13}$$

where

$$\begin{aligned}
\mathbf{u}_x &= \frac{[\mathbf{R}_{LSI}(t_{LAND})]}{|\mathbf{R}_{LSI}(t_{LAND})|},\\
\mathbf{u}_z &= \frac{[(\mathbf{R}_I(t_I) \times \mathbf{V}_I(t_I)) \times \mathbf{u}_x]}{|(\mathbf{R}_I(t_I) \times \mathbf{V}_I(t_I)) \times \mathbf{u}_x|},\\
\mathbf{u}_y &= \mathbf{u}_z \times \mathbf{u}_x.
\end{aligned}$$

Consider the following notations:

$$\begin{aligned}
\mathbf{T}_I^P &= [a_{ij}] = \sum_{k=1}^{3} c_{ik} b_{kj},\\
\mathbf{T}_I^L(t_{LAND}) &= [b_{ij}],\\
\mathbf{T}_L^P(t_{LAND}) &= [c_{ij}], \qquad i, j = 1, 2, 3,
\end{aligned}$$

$$\begin{aligned}
\mathbf{h}_I &= [\mathbf{R}_I(t_I) \times \mathbf{V}_I(t_I)] = (h_x, h_y, h_z),\\
\mathbf{u}_x &= (a_{11}, a_{12}, a_{13}),\\
\mathbf{u}_y &= (a_{21}, a_{22}, a_{23}),\\
\mathbf{u}_z &= (a_{31}, a_{32}, a_{33}),
\end{aligned}$$

It is assumed that $\mathbf{T}_I^L(t_{LAND})$ is completely determined prior to calculations of $\mathbf{T}_I^P$. Then

$$\mathbf{R}_{LSI}(t_{LAND}) = \mathbf{T}_L^I(t_{LAND})\mathbf{R}_{LSL}. \tag{4.14}$$

Note that $\mathbf{T}_I^P$ can be computed in the form:

$$\mathbf{T}_I^P = \mathbf{T}_L^P(t_{LAND})\mathbf{T}_I^L(t_{LAND}). \tag{4.15}$$

References

1. Project Apollo Coordinate System Standards, Apollo Program, SE 008-001-1, NASA Office of Manned Space Flight (1965). (See http://ntrs.nasa.gov/archive/nasa/casi.ntrs.nasa.gov/19700076120_1970076120.pdf)
2. “Guidance System Operations Plan For Manned LM Earth Orbital and Lunar Missions Using Program Luminary 1C (LM131 Rev. 1),” Section 5. Guidance Equations. (Rev.8). Apollo Guidance Navigation Control, R-567, MIT, Cambridge, MA (1970)
3. Report of the IAU/IAG Working Group on Cartographic Coordinates and Rotational Elements of the Planets and Satellites. (See http://astrogeology.usgs.gov/Projects/ISPRS/PREPRINTS/index_preprints.html)

Chapter 5
Approach-Phase Ground-Based Targeting

5.1 Approach-Phase Targeting Equations

5.1.1 Preliminary Expressions for the State Vector

Let's define the state vector **X** of the lunar module (LM) as $\mathbf{X} = (\mathbf{R}, \mathbf{V}, \mathbf{A}, \mathbf{J}, \mathbf{S})$. This definition of the state vector will be used throughout the monograph. Time interval between terminal instant and middle points of the approach phase is computed as [1]

$$t_f = TAPM - TAPF.$$

Duration of the landing maneuver on this phase is determined via target referenced times as

$$t_i = TAPI - TAPF.$$

One can also compute the following position and velocity components according to the constraint about $\gamma = 16$ degrees trajectory slope on the approach phase:

$$R_{ZAPMG} = -R_{XAPMG}/tan(\gamma),$$

$$V_{ZAPMG} = -V_{XAPMG}/tan(\gamma),$$

$$R_{ZAPIG} = -R_{XAPIG}/tan(\gamma).$$

Using the definition of a quartic:

© The Author(s), under exclusive license to Springer Nature Switzerland AG 2026
D. M. Azimov and R. H. Bishop, *Apollo-Class Targeting and Guidance for Lunar Powered Descent and Precision Landing*, Springer Aerospace Technology,
https://doi.org/10.1007/978-3-031-91088-3_5

$$\begin{bmatrix} R_{XAPMG} \\ V_{XAPMG} \\ R_{XAPIG} \end{bmatrix} = \begin{bmatrix} 1 & t_f & \frac{t_f^2}{2} & \frac{t_f^3}{6} & \frac{t_f^4}{24} \\ 0 & 1 & t_f & \frac{t_f^2}{2} & \frac{t_f^3}{6} \\ 1 & t_i & \frac{t_i^2}{2} & \frac{t_i^3}{6} & \frac{t_i^4}{24} \end{bmatrix} \begin{bmatrix} R_{XAPFG} \\ V_{XAPFG} \\ A_{XAPFG} \\ J_{XAPFG} \\ S_{XAPFG} \end{bmatrix},$$

or

$$R_{XAPMG} - R_{XAPFG} - V_{XAPFG} t_f = A_{XAPFG} \frac{t_f^2}{2} + J_{XAPFG} \frac{t_f^3}{6} + S_{XAPFG} \frac{t_f^4}{24},$$

$$V_{XAPMG} - 0 - V_{XAPFG} = A_{XAPFG} t_f + J_{XAPFG} \frac{t_f^2}{2} + S_{XAPFG} \frac{t_f^3}{6},$$

$$R_{XAPIG} - R_{XAPFG} - V_{XAPFG} t_i = A_{XAPFG} \frac{t_i^2}{2} + J_{XAPFG} \frac{t_i^3}{6} + S_{XAPFG} \frac{t_i^4}{24}.$$

and by rearranging, one can rewrite this equation as

$$\begin{bmatrix} \frac{t_f^2}{2} & \frac{t_f^3}{6} & \frac{t_f^4}{24} \\ t_f & \frac{t_f^2}{2} & \frac{t_f^3}{6} \\ \frac{t_i^2}{2} & \frac{t_i^3}{6} & \frac{t_i^4}{24} \end{bmatrix} \begin{bmatrix} A_{XAPFG} \\ J_{XAPFG} \\ S_{XAPFG} \end{bmatrix} = \begin{bmatrix} -1 & -t_f & 1 & 0 & 0 \\ 0 & -1 & 0 & 1 & 0 \\ -1 & -t_i & 0 & 0 & 1 \end{bmatrix} \begin{bmatrix} R_{XAPFG} \\ V_{XAPFG} \\ R_{XAPMG} \\ V_{XAPMG} \\ R_{XAPIG} \end{bmatrix},$$

Terminal X components:

$$\begin{bmatrix} A_{XAPFG} \\ J_{XAPFG} \\ S_{XAPFG} \end{bmatrix} = \begin{bmatrix} \frac{t_f^2}{2} & \frac{t_f^3}{6} & \frac{t_f^4}{24} \\ t_f & \frac{t_f^2}{2} & \frac{t_f^3}{6} \\ \frac{t_i^2}{2} & \frac{t_i^3}{6} & \frac{t_i^4}{24} \end{bmatrix}^{-1} \begin{bmatrix} -1 & -t_f & 1 & 0 & 0 \\ 0 & -1 & 0 & 1 & 0 \\ -1 & -t_i & 0 & 0 & 1 \end{bmatrix} \begin{bmatrix} R_{XAPFG} \\ V_{XAPFG} \\ R_{XAPMG} \\ V_{XAPMG} \\ R_{XAPIG} \end{bmatrix}.$$

The trajectory of the maneuver is confined to the XZ-plane, and therefore the terminal Y-components of the state are all zeros:

$$R_{YAPFG} = 0, \quad V_{YAPFG} = 0, \quad A_{YAPFG} = 0, \quad J_{YAPFG} = 0, \quad S_{YAPFG} = 0.$$

One can obtain an equation for the Z-components using the formula:

$$\begin{bmatrix} R_{ZAPMG} \\ V_{ZAPMG} \\ R_{ZAPIG} \end{bmatrix} = \begin{bmatrix} 1 & t_f & \frac{t_f^2}{2} & \frac{t_f^3}{6} & \frac{t_f^4}{24} \\ 0 & 1 & t_f & \frac{t_f^2}{2} & \frac{t_f^3}{6} \\ 1 & t_i & \frac{t_i^2}{2} & \frac{t_i^3}{6} & \frac{t_i^4}{24} \end{bmatrix} \begin{bmatrix} R_{ZAPFG} \\ V_{ZAPFG} \\ A_{ZAPFG} \\ J_{ZAPFG} \\ S_{ZAPFG} \end{bmatrix},$$

The remaining components of position and velocity vectors, R_{ZAPFG} and V_{ZAPFG} at the AP- terminal point are determined based on the constraints at this point. This point is also considered as the initial point of the terminal descent phase. These constraints, called as the P66 compatibility constraints, are imposed to null the Z components of position and velocity vectors, R_Z and V_Z shortly after the terminal descent phase starts. Assuming that the trajectory on approach phase is quartic and the Z components of the jerk and snap vectors are zeros, $J_Z = 0$ and $S_Z = 0$ on this phase, i.e. constant total acceleration, $A_Z = constant$, one can compute that at any given time, t,

$$R_Z(t) = R_{ZAPFG} + V_{ZAPFG}t + A_{ZAPFG}\frac{t^2}{2}$$

and

$$V_Z(t) = V_{ZAPFG} + A_{ZAPFG}t.$$

If at some time $t = \tau$ on terminal descent phase these components reach zeros, i.e.

$$R_Z(\tau) = 0, \qquad V_Z(\tau) = 0,$$

then one can find that

$$R_{ZAPFG} = -V_{ZAPFG}\tau - A_{ZAPFG}\frac{\tau^2}{2}$$

or

$$R_{ZAPFG} = -V_{ZAPFG}\frac{\tau}{2},$$

and

$$V_{ZAPFG} = -A_{ZAPFG}\tau,$$

where τ is the effective P66 horizontal time constant. For example, $\tau = 8$ s was found satisfactory in the P66 algorithm [1]. Note that in this reference, the expression for R_{ZAPFG} has no denominator. The latter two formulas represent the P66 constraints for R_{ZAPFG} and V_{ZAPFG} imposed at the final point of the approach phase.

Using these constraints, from the equation for the Z-components it follows that

$$\begin{bmatrix} A_{ZAPFG} \\ J_{ZAPFG} \\ S_{ZAPFG} \end{bmatrix} = \begin{bmatrix} \frac{\tau^2}{2} - \tau t_f + \frac{t_f^2}{2} & \frac{t_f^3}{6} & \frac{t_f^4}{24} \\ 0 - \tau + t_f & \frac{t_f^2}{2} & \frac{t_f^3}{6} \\ \frac{\tau^2}{2} - \tau t_i + \frac{t_i^2}{2} & \frac{t_i^3}{6} & \frac{t_i^4}{24} \end{bmatrix}^{-1} \begin{bmatrix} R_{ZAPMG} \\ V_{ZAPMG} \\ R_{ZAPIG} \end{bmatrix}.$$

5.1.2 *The Approach-Phase Target, Initial and Terminal States*

The complete target state can now be computed via a forward propagation from $TAPF$ to 0 as follows [1, 2]:

$$\mathbf{X}_{APTG} = STM(0, TAPF)\mathbf{X}_{APFG},$$

where the state transformation matrix $STM(0, TAPF)$ is calculated based on the general formula:

$$STM(t_1, t_0) = \begin{bmatrix} 1 & (t_1 - t_0) & \frac{(t_1-t_0)^2}{2} & \frac{(t_1-t_0)^3}{6} & \frac{(t_1-t_0)^4}{24} \\ 0 & 1 & (t_1 - t_0) & \frac{(t_1-t_0)^2}{2} & \frac{(t_1-t_0)^3}{6} \\ 0 & 0 & 1 & (t_1 - t_0) & \frac{(t_1-t_0)^2}{2} \\ 0 & 0 & 0 & 1 & (t_1 - t_0) \\ 0 & 0 & 0 & 0 & 1 \end{bmatrix}. \tag{5.1}$$

As the approach phase is represented as a quartic trajectory, the complete initial state can be found by a backward propagation from the target point:

$$\mathbf{X}_{APIG} = STM(TAPI, 0)\mathbf{X}_{APTG}.$$

As it can be seen from the above expressions, the AP trajectory can be determined in a closed form.

5.2 Approach-Phase Targeting Constraints, Inputs and Outputs

As was shown in Ref. [1], a quartic polynomial is of a minimum order with which five constraints, namely, the constraints imposed on the position, velocity, acceleration, jerk and snap vectors on the reference trajectory are satisfied. This allowed for an approximation of the AP-trajectory as a quartic polynomial and therefore, the AP-

Table 5.1 AP-targeting constraints

The AP terminal point, $TAPF$:	$R_{XAPFG} = 30\ m,\ V_{XAPFG} = -1$ m/s $R_{ZAPFG} = A_{ZAPFG}\tau^2/2,$ $V_{ZAPFG} = -A_{ZAPFG}\tau$
The AP middle point, $TAPM$:	$R_{XAPMG} = 150.0$ m, $V_{XAPMG} = -5.0$ m/s
The AP initial point, $TAPI$:	$R_{XAPIG} = 2200.0$ m

targeting solutions are in a closed form. The AP-targeting algorithm is executed several times as part of the targeting program. Note that the differences of the AP-targeting algorithm described below from the Apollo targeting algorithms are that in the former algorithm, R_{ZAPFG} has more accurate expression and the order of its execution, together with BP-targeting, provides more accurate target states. The AP-targeting constraints need to be satisfied at the initial, middle and terminal points of the phase and are used to determine the AP-quartic in a closed form (see Fig.3.2 of Chap. 3). Also, the values of the constraint parameters are comparable to those of the original Apollo program. The AP-targeting constraints, inputs and outputs are given in Tables 5.1, 5.2 and 5.3.

5.2.1 *Constraints*

The AP-targeting constraints are given in Table 5.1 and the P66 is the terminal-descent-phase guidance algorithm which controls horizontal and vertical velocities on this phase. According to Klumpp, $\tau = 8$ s was found satisfactory in the AP targeting analysis [1]. The remaining components of the position and velocity vectors are computed as follows ($\gamma = 16°$):

$$R_{ZAPMG} = -R_{XAPMG}/tan(\gamma),\ V_{ZAPMG} = -V_{XAPMG}/tan(\gamma),$$

$$R_{ZAPIG} = -R_{XAPIG}/tan(\gamma),\ R_{YAPIG} = R_{YAPMG} = R_{YAPFG} = 0,$$

$$V_{YAPIG} = V_{YAPMG} = V_{YAPFG} = 0.$$

5.2.2 *Inputs and Outputs*

The AP-targeting algorithm inputs are given in Table 5.2. The AP-targeting algorithm outputs consist of the complete target and initial states are given in Table 5.3.

Table 5.2 AP-targeting inputs

Terminal, midpoint and initial target referenced times	$TAPE = -10$ s, $TAPM = -76$ s $TAPI = -176$ s
The landing site must be approached with this slope	$\gamma = 16$ deg
t_{LAND} and effective P66 horizontal time constant	$t_{LAND} = 1320$ s and $\tau = 8$ s
Terminal altitude and altitude rate	$R_{XAPFG} = 30$ m, $V_{XAPFG} = -1$ m/s
The X-components of position and velocity at midpoint	$R_{XAPMG} = 150.0$ m $V_{XAPMG} = -5.0$ m/s
The X-component of the initial position	$R_{XAPIG} = 2220.0$ m

Table 5.3 AP-targeting outputs

Target states	$\mathbf{X}_{APTG} = (\mathbf{R}_{APTG}\ \mathbf{V}_{APTG}, \mathbf{A}_{APTG}, \mathbf{J}_{APTG}, \mathbf{S}_{APTG})$
Initial states	$\mathbf{X}_{APIG} = (\mathbf{R}_{APIG}\ \mathbf{V}_{APIG}, \mathbf{A}_{APIG}, \mathbf{J}_{APIG}, \mathbf{S}_{APIG})$

5.3 Approach-Phase Targeting Algorithm

(1) Compute the time interval between terminal and middle points, and the duration of the phase:

$$t_f = TAPM - TAPF, \qquad t_i = TAPI - TAPF.$$

(2) Compute the position and velocity components according to the constraints:

$$R_{ZAPMG} = -R_{XAPMG}/tan(\gamma),\ V_{ZAPMG} = -V_{XAPMG}/tan(\gamma),$$

$$R_{ZAPIG} = -R_{XAPIG}/tan(\gamma).$$

(3) Set the position and velocity components on Y-axis (out-of-plane) to zero:

$$R_{YAPIG} = R_{YAPMG} = R_{YAPFG} = 0, \quad V_{YAPIG} = V_{YAPMG} = V_{YAPFG} = 0.$$

(4) Compute the terminal X-components:

$$\begin{bmatrix} A_{XAPFG} \\ J_{XAPFG} \\ S_{XAPFG} \end{bmatrix} = \begin{bmatrix} \frac{t_f^2}{2} & \frac{t_f^3}{6} & \frac{t_f^4}{24} \\ t_f & \frac{t_f^2}{2} & \frac{t_f^3}{6} \\ \frac{t_i^2}{2} & \frac{t_i^3}{6} & \frac{t_i^4}{24} \end{bmatrix}^{-1} \begin{bmatrix} -1 & -t_f & 1 & 0 & 0 \\ 0 & -1 & 0 & 1 & 0 \\ -1 & -t_i & 0 & 0 & 1 \end{bmatrix} \begin{bmatrix} R_{XAPFG} \\ V_{XAPFG} \\ R_{XAPMG} \\ V_{XAPMG} \\ R_{XAPIG} \end{bmatrix}.$$

(5) Determine the terminal Y-components:

$$R_{YAPFG} = 0, \quad V_{YAPFG} = 0, \quad A_{YAPFG} = 0, \quad J_{YAPFG} = 0, \quad S_{YAPFG} = 0.$$

(6) Compute the Z components of the acceleration, jerk and snap vectors:

$$\begin{bmatrix} A_{ZAPFG} \\ J_{ZAPFG} \\ S_{ZAPFG} \end{bmatrix} = \begin{bmatrix} \frac{\tau^2}{2} - \tau t_f + \frac{t_f^2}{2} & \frac{t_f^3}{6} & \frac{t_f^4}{24} \\ 0 - \tau + t_f & \frac{t_f^2}{2} & \frac{t_f^3}{6} \\ \frac{\tau^2}{2} - \tau t_i + \frac{t_i^2}{2} & \frac{t_i^3}{6} & \frac{t_i^4}{24} \end{bmatrix}^{-1} \begin{bmatrix} R_{ZAPMG} \\ V_{ZAPMG} \\ R_{ZAPIG} \end{bmatrix}.$$

(7) Compute the Z-components of the position and velocity vectors:

$$R_{ZAPFG} = -V_{ZAPFG}\frac{\tau}{2}, \qquad V_{ZAPFG} = -A_{ZAPFG}\tau.$$

(8) Compute $STM(0, TAPF)$ and $STM(TAPI, 0)$ using the general formula:

$$STM(t_1, t_0) = \begin{bmatrix} 1 & (t_1 - t_0) & \frac{(t_1-t_0)^2}{2} & \frac{(t_1-t_0)^3}{6} & \frac{(t_1-t_0)^4}{24} \\ 0 & 1 & (t_1 - t_0) & \frac{(t_1-t_0)^2}{2} & \frac{(t_1-t_0)^3}{6} \\ 0 & 0 & 1 & (t_1 - t_0) & \frac{(t_1-t_0)^2}{2} \\ 0 & 0 & 0 & 1 & (t_1 - t_0) \\ 0 & 0 & 0 & 0 & 1 \end{bmatrix}.$$

(9) Compute the complete target and initial states (Fig. 5.1):

$$\mathbf{X}_{APTG} = STM(0, TAPF)\mathbf{X}_{APFG},$$

$$\mathbf{X}_{APIG} = STM(TAPI, 0)\mathbf{X}_{APTG}.$$

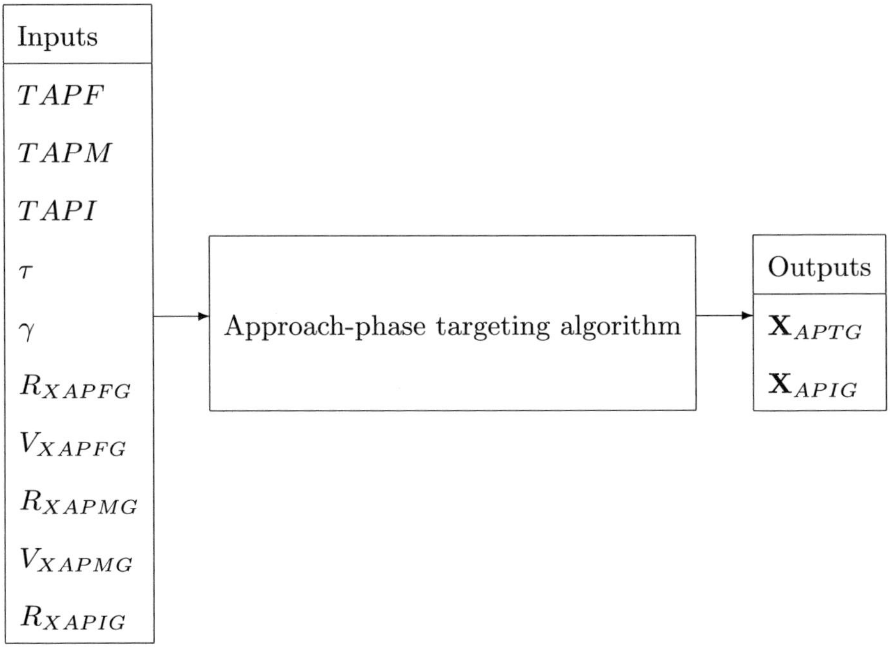

Fig. 5.1 Schematic illustration of the approach-phase targeting procedure

References

1. Klumpp AR (1971) "Apollo Lunar-Descent Guidance," Apollo Guidance Navigation Control, R-695. Massachusetts Institute of Technology, MIT Instrumentation Laboratory Report, Cambridge, MA
2. Klumpp AR (1968) Manually retargeted automatic landing system for the Lunar Module (LM). AIAA J Spacecraft Rockets 5(2):129–138

Chapter 6
Braking-Phase Targeting

6.1 Introductory Remarks

Determination of the BP-quartic parameters represents an iterative process because the constraints of the quartic do not uniquely determine these parameters in a closed-form. This process, in general, involves the AP-targeting procedures in order to provide a state continuity at the point where both BP- and AP-quartics are connected (see Fig. 4.2 of Chap. 4). Note that no closed-form solution was used to determine the pre-quartic segment of the BP-trajectory with constant thrust (CT) arc (BP-CT-arc) [1]. The list of inputs includes more parameters than the Apollo program and in general, is the same for the targeting procedures. Unlike the existing descriptions of the Apollo targeting algorithm, the BP-targeting algorithm given below includes all necessary computations. Each computation is explicit in terms of all procedures and consistent with the inputs and outputs of the ignition and guidance algorithms, and provides desirable targeting results that are in good agreement with post-flight data.

6.2 Time and Mass Estimations

The time remaining to land from the BP-terminal point, BRF is computed as

$$\Delta t_{LBRF} = \Delta t_{LTDI} + |TAPI - TAPF|, \tag{6.1}$$

where Δt_{LTDI} is the estimated time remaining to land from the AP-terminal point. The time interval between the TCR point and landing is

$$\Delta t_{LTHR} = \Delta t_{LBRF} + \Delta t_{THR}. \tag{6.2}$$

Consequently, the duration of the BP-CT-arc can be determined as

© The Author(s), under exclusive license to Springer Nature Switzerland AG 2026

D. M. Azimov and R. H. Bishop, *Apollo-Class Targeting and Guidance for Lunar Powered Descent and Precision Landing*, Springer Aerospace Technology,
https://doi.org/10.1007/978-3-031-91088-3_6

$$\Delta t_{BRF} = t_{LAND} - t_I - \Delta t_{LTHR}, \quad t_I = t_{IG} + \Delta t_{TRIM}, \quad t_{IG} = t_0 + \Delta t_{BA}, \tag{6.3}$$

where t_0 is the clock-time at which BP is initiated by keyboard entry about 10 min before the nominal ignition time, Δt_{BA} is the time interval between t_0 and t_{IG}, and t_{LAND} is the input to AP-targeting algorithm.

The initial time t_0 of the landing maneuver can be determined as follows. It will be assumed that the radius of the initial orbit is the same as the apolune r_a of the entry orbit. If r_p is the perilune of the entry orbit, then the semi-major axis, a_{trans} of this orbit is computed using

$$a_{trans} = \frac{r_p + r_a}{2}, \qquad r_p = R_M + h_p, \qquad r_a = R_M + h_a. \tag{6.4}$$

If the lander is transferred from the initial orbit into the entry orbit at its apolune with $|\mathbf{R}_I(t_x)| = r_a$ by means of a finite thrust or an impulse, and if BP-CT-arc starts at the perilune of this orbit at t_I, that is $|\mathbf{R}_I(t_I)| = r_p$, then the entry orbit can be considered as a Hohmann transfer orbit with time of flight

$$\Delta t_h = \pi\sqrt{\frac{a_{trans}^3}{\mu_M}}. \tag{6.5}$$

In this case $t_I = t_x + \Delta t_h$. Note that although t_x is a mission dependent parameter, in the BP-algorithm it is set to 0 without loss of generality. As the ballistic part of BP is started at t_0 and followed by the ullage and trim phases, then

$$t_0 = t_x + [\pi\sqrt{\frac{a_{trans}^3}{\mu_M}} - \Delta t_{BA} - \Delta t_{TRIM}]. \tag{6.6}$$

This means that once the elapsed clock time t_x at the apolune of the entry orbit is given, the initial time t_0 of the landing maneuver can be estimated by Eq. (6.6).

In order to estimate the mass at the BP-terminal point, one can compute the mass estimates for each part of the trajectory until this point. The mass estimate at final instant of the trim phase is computed using the velocity increment:

$$\Delta V_{TRIMF} = A_{TRIM}\Delta t_{TRIM}, \qquad \Delta M_{TRIM} = \Delta V_{TRIMF}\frac{M_0}{V_E}. \tag{6.7}$$

So, the mass estimate after the trim phase is given as

$$M_{TRIMF} = M_0 - \Delta M_{TRIM}. \tag{6.8}$$

Determination of the thrust vector magnitude is based on a thrust constraint that on BP-CT-arc and quartic the lander utilizes the 93% and 57% of DPS maximum thrust respectively. Defining them as F_{THR} and F_{BRF}, one can have

$$F_{THR} = 0.93F_{max}, \quad F_{BRF} = 0.57F_{max}, \quad F_{max} = V_E\dot{M}_{DPS}, \tag{6.9}$$
$$\dot{M}_{THR} = \frac{F_{THR}}{V_E}, \quad \dot{M}_{BRF} = \frac{F_{BRF}}{V_E}.$$

The mass estimates for the BP-CT-arc and BP-quartic can now be determined as:

$$M_{THR} = M_{TRIMF} - \dot{M}_{THR}\Delta t_{BRF}, \quad M_{BRF} = M_{THR} - \dot{M}_{BRF}\Delta t_{THR}. \tag{6.10}$$

6.3 Estimation of Terminal Jerk and Snap Vectors for First Iteration

The components of the jerk vector at BP-terminal point are computed in such a way to produce zero rate of change of thrust at this point. It can be shown that at this point one can have

$$\mathbf{A}_{BRFG} = \left(\frac{F_{BRFG}}{M_{BRF}}\right)\mathbf{U}_{BRFG} - \mathbf{G}_{MG}(G_0,\ 0,\ 0), \tag{6.11}$$

$$\mathbf{J}_{BRFG} = \dot{\mathbf{A}}_{BRFG} = \left(-(A_{XBRFG} + G_M)\frac{\dot{M}_{BRF}}{M_{BRF}},\ 0,\ -A_{ZBRFG}\frac{\dot{M}_{BRF}}{M_{BRF}}\right). \tag{6.12}$$

$$\mathbf{S}_{BRFG} = \dot{\mathbf{J}}_{BRFG} = \left(-2J_{XBRFG}\frac{\dot{M}_{BRF}}{M_{BRF}},\ 0,\ -2J_{ZBRFG}\frac{\dot{M}_{BRF}}{M_{BRF}}\right), \tag{6.13}$$

where $\dot{F}_{BRF} = 0$. It should be noted that for comparison purposes, in Apollo BP-targeting program the terminal jerk and snap vectors have the following form for the first iteration:

$$\mathbf{J}_{BRFG} = \left(0,\ 0,\ -A_{ZBRFG}\frac{\dot{M}_{BRF}}{M_{BRF}}\right), \qquad \mathbf{S}_{BRFG} = (0,\ 0,\ 0). \tag{6.14}$$

If these vectors are compared to the vectors given in Eqs. (6.12) and (6.13), then one can find that in Eq. (6.14) of Apollo targeting program, it is accepted that $A_{XBRFG} + G_M = 0$ and $S_{XBRFG} = 0$. Equations (6.12) and (6.13) are obtained by direct differentiation of $\mathbf{A}_{BRFG}$ and $\mathbf{J}_{BRFG}$ without any simplifications and therefore, it is recommended to use the vectors given in these equations for the first iteration in the BP-targeting program.

6.4 Expressions for Jerk and Snap Vectors

As is known, at any given target referenced time T, the position and velocity vectors, $\mathbf{R}$, $\mathbf{V}$ of a point on a quartic in any given coordinate system can be expressed by the following formulas [1]:

$$\begin{bmatrix} \mathbf{R} \\ \mathbf{V} \end{bmatrix} = \begin{bmatrix} 1 & T & \frac{T^2}{2} & \frac{T^3}{6} & \frac{T^4}{24} \\ 0 & 1 & T & \frac{T^2}{2} & \frac{T^3}{6} \end{bmatrix} \begin{bmatrix} \mathbf{R}_T \\ \mathbf{V}_T \\ \mathbf{A}_T \\ \mathbf{J}_T \\ \mathbf{S}_T \end{bmatrix},$$

where T is target referenced time, $\mathbf{R}_T, \mathbf{V}_T, \mathbf{A}_T, \mathbf{J}_T, \mathbf{S}_T$ are the position, velocity, acceleration, jerk and snap vectors respectively at time T. This formulas can be rewritten as

$$\mathbf{R} = \begin{bmatrix} 1 & T & \frac{T^2}{2} \end{bmatrix} \begin{bmatrix} \mathbf{R}_T \\ \mathbf{V}_T \\ \mathbf{A}_T \end{bmatrix} + \frac{T^3}{6}\mathbf{J}_T + \frac{T^4}{24}\mathbf{S}_T,$$

$$\mathbf{V} = \begin{bmatrix} 0 & 1 & T \end{bmatrix} \begin{bmatrix} \mathbf{R}_T \\ \mathbf{V}_T \\ \mathbf{A}_T \end{bmatrix} + \frac{T^2}{2}\mathbf{J}_T + \frac{T^3}{6}\mathbf{S}_T.$$

These two expressions can be rearranged to find the jerk and snap vectors in the form:

$$\begin{bmatrix} \mathbf{J}_T \\ \mathbf{S}_T \end{bmatrix} = \begin{bmatrix} -\frac{24}{T^3} & -\frac{18}{T^2} & -\frac{6}{T} & \frac{24}{T^3} & -\frac{6}{T^2} \\ \frac{72}{T^4} & \frac{48}{T^3} & \frac{12}{T^2} & -\frac{72}{T^4} & \frac{24}{T^3} \end{bmatrix} \begin{bmatrix} \mathbf{R}_T \\ \mathbf{V}_T \\ \mathbf{A}_T \\ \mathbf{R} \\ \mathbf{V} \end{bmatrix}.$$

6.5 Braking-Phase Targeting Constraints, Inputs and Outputs

As it was mentioned in the Klumpp's document, there was no closed form solution to determine the BR quartic as the number of constraints is insufficient to determine all states at initial and terminal points [1]. There exist seven constraints, specified arbitrarily, and three conditions should be found by iteration. These constraints, inputs and outputs given below.

6.5.1 Terminal State Constraints

$$\mathbf{R}_{BRFG} = \mathbf{R}_{APIG},$$
$$\mathbf{V}_{BRFG} = \mathbf{V}_{APIG},$$
$$\mathbf{A}_{BRFG} = \left(\frac{F_{BRF}}{M_{BRF}}\mathbf{U}_{BRFG} - \mathbf{G}_{MG}(G_0,\ 0,\ 0)\right),$$
$$\mathbf{J}_{BRFG} = \left(-(A_{XBRFG} + G_M)\frac{\dot{M}_{BRF}}{M_{BRF}},\ 0,\ -A_{ZBRFG}\frac{\dot{M}_{BRF}}{M_{BRF}}\right),$$
$$Y_{BRFG} = 0,$$
$$\theta_{PBRF} = 60 \text{ deg}.$$

6.5.2 Operational Constraints

$\Delta t_{THR} = 120$ s of motion with $57 \div 63$ percent thrust during the throttle recovery, starting from TDP ("throttle down" point) and ending at BRF ("high gate") [2]:

$$F_{THR} = 0.93\ F_{MAX}, \qquad F_{BRF} = 0.57\ F_{MAX},$$

6.5.3 Inputs and Outputs

The BP-targeting algorithm inputs are given in Table 6.1. The BP-targeting algorithm outputs are given in Table 6.2. These outputs consist of the complete target state, and position and velocity vectors.

6.6 Braking-Phase Targeting Algorithm

The flow-chart of the BP algorithm provided in Fig. 6.1 in two parts is helpful to understand the overall targeting procedure. Detailed and itemized form of the algorithm is provided below:

(1) Set the Moon mean radius, R_M. For simulation purposes, one can choose $R_M = (R_e + R_p)/2$. This quantity can be computed using one of the following options for the lunar surface:
– sphere with mean radius of the Moon, R_M;
– reference ellipsoid with polar and equatorial radii of the Moon;

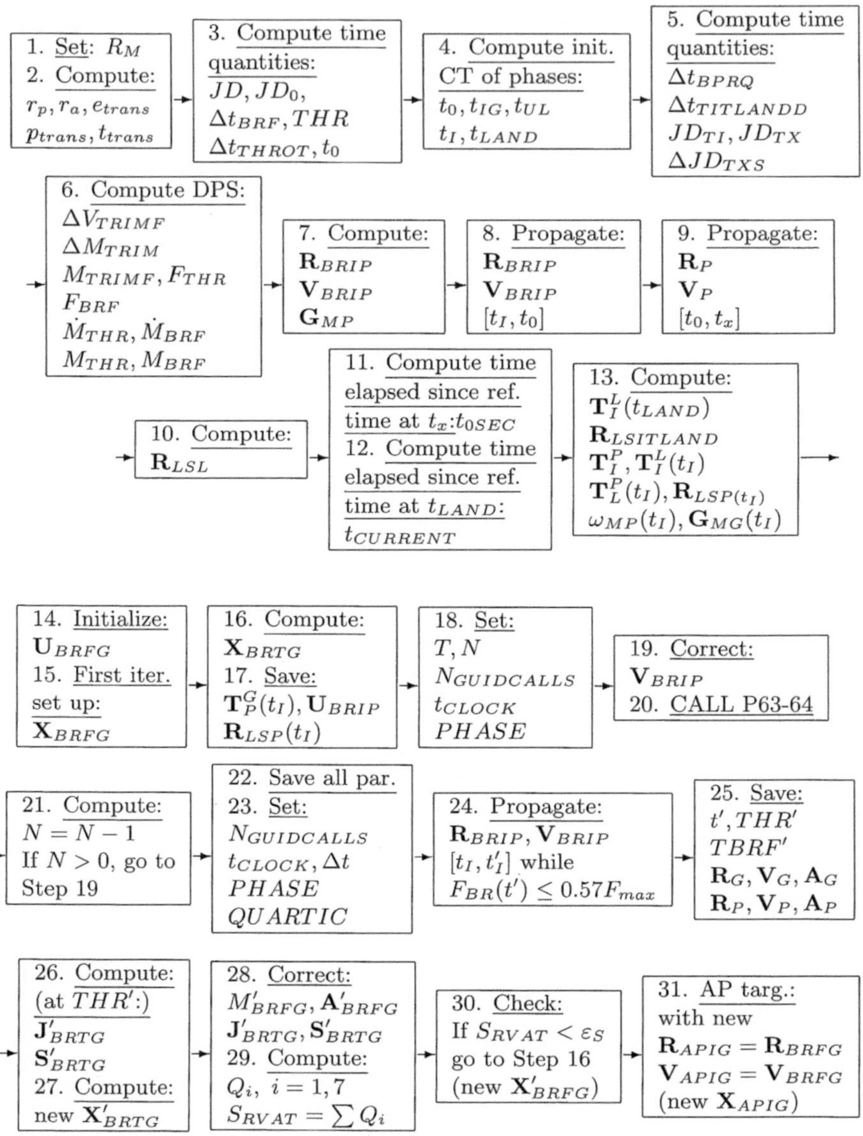

Fig. 6.1 Flow-chart of the BP algorithm

Table 6.1 Inputs of BP of Apollo 11

Moon's gravitational acceleration	$G_0 = 1.62$ m/s^2
Moon mean radius	$R_M = 1736500$ m
LS position vector magnitude, latitude, longitude	H_L, φ_L, θ_L
Altitudes of perilune and apolune of entry orbit	$h_p = 15000$, $h_a = 120000$ m
Angle between LS and lander's position vectors at t_0 measured in $X_P Z_P$-plane	$\theta_0 = 16°$
P64 initial state	$\mathbf{R}_{APIG}, \mathbf{V}_{APIG}$
Terminal pitch angle constraint	$\theta_{PBRF} = 60.0$ deg
Duration of trim phase starting from t_{IG} to t_I	$\Delta t_{TRIM} = 26$ s
Duration of ballistic phase starting from t_0 to t_{IG}	$\Delta t_{BA} = 600$ s
Duration of ullage	$\Delta t_{ULLAGE} = 7.5$ s
Time interval between TCR and nominal t_{LAND}	$\Delta t_{LTDI} = 31$ s
Required time between TCR and BP-terminal point	$\Delta t_{THR} = 120$ s
Nominal landing time counted from t_0	$t_{LAND} = 1320$ s
Initial target referenced time at first guidance call at t_I	$T = -664$ s
Target referenced time at TCR	$THR = -180$ s
Magnitude of thrust acceleration on trim phase	$\lvert\mathbf{A}_{TRIM}\rvert = 0.35$ m/s^2
Initial mass of the lander	$M_0 = 15079$ kg
Magnitude of exhaust velocity of gas particles in DPS	$V_E = 3150$ m/s
DPS maximum thrust	$F_{max} = 46706$ N
Time elapsed between July 1, 1969 and t_x (mission dep.)	$t_{0DAYS} = 20$ days
Convergence criterion for relative errors in position, velocity, acceleration	$\varepsilon_S = 10^{-4}$

Table 6.2 Outputs of BP of Apollo 11

Target state	$\mathbf{X}_T = (\mathbf{R}_{BRTG}, \mathbf{V}_{BRTG}, \mathbf{A}_{BRTG}, \mathbf{J}_{BRTG}, \mathbf{S}_{BRTG})$
Nominal initial position	$\mathbf{R}_{BRIG}$
Nominal initial velocity	$\mathbf{V}_{BRIG}$

– topographical surface model represented by the topographic data tables;
– surface model represented by high order spherical harmonics.

(2) Compute perilune and apolune of the entry orbit:

$$r_p = R_M + h_p, \quad r_a = R_M + h_a.$$

(3) Compute semi-major axis, eccentricity and semi latus rectum of the entry orbit:

$$a_{trans} = \frac{r_a + r_p}{2}, \quad e_{trans} = \frac{r_a - r_p}{2a_{trans}}, \quad p_{trans} = a_{trans}(1 - e_{trans}^2).$$

(4) Compute the Julian date of given nominal landing time including the time zone:

$$(Year, month, date, hour, minute, second, timezone) \Rightarrow JD_{LAND}.$$

(5) Compute the Julian date of July 1, 1969, at (0:0:0.0) including the time zone:

$$(1969, 7, 1, 0, 0, 0.0, time\ zone) \Rightarrow JD_0.$$

(6) Compute the time to land from the BP-terminal point:

$$\Delta t_{LBRF} = \Delta t_{LTDI} + |TAPI - TAPF|.$$

(7) Compute target referenced time at throttle control recovery (TCR) point:

$$THR = -(\Delta t_{THR} + |TBRF|).$$

(8) Compute the time interval between the throttle recovery point and landing instant:

$$\Delta t_{LTHR} = \Delta t_{LBRF} + \Delta t_{THR}.$$

(9) Compute the transfer time from apolune to perilune of the entry orbit:

$$t_{trans} = \pi\sqrt{a_{trans}^3/\mu_M}.$$

(10) Compute the initial clock time of the maneuver (start P63):

$$t_0 = t_{trans} - \Delta t_{BA} - \Delta t_{TRIM}.$$

(11) Compute the nominal clock time of ignition:

$$t_{IG} = t_0 + \Delta t_{BA}.$$

(12) Compute the clock time of ullage:

$$t_{ULLAGE} = t_{IG} - \Delta t_{ULLAGE}.$$

(13) Compute the nominal initial time of the CT segment (clock time of the first guidance pass):

$$t_I = t_{IG} + \Delta t_{TRIM}.$$

(14) Compute the nominal clock time of nominal landing (since t_x):

$$t_{LAND} = t_0 + \Delta t_{LAND}.$$

(15) Compute the duration of the BP-CT-arc:

$$\Delta t_{BRPQ} = t_{LAND} - t_I - \Delta t_{LTHR}.$$

(16) Compute the time duration from t_I to t_{LAND}:

$$\Delta t_{TITLANDD} = 1.1574 \times 10^{-5}(t_{LAND} - t_I).$$

(17) Compute the time in Julian dates at perilune:

$$JD_{TI} = JD_{LAND} - \Delta_{TITLANDD}.$$

(18) Compute the time in Julian dates at apolune (at t_x):

$$JD_{TX} = JD_{TI} - 1.1574 \times 10^{-5} t_{trans}.$$

(19) Compute the time elapsed since July 1, 1969 in Julian dates at apolune (at t_x) and convert to seconds:

$$\Delta JD_{TXS} = 86400(JD_{TX} - JD_0).$$

(20) Compute the time elapsed since July 1, 1969 in Julian dates at t_{LAND} and convert to seconds:

$$\Delta JD_{0JDLANDS} = 86400(JD_{LAND} - JD_0).$$

(21) Compute the time elapsed since July 1, 1969 in Julian dates at t_I and convert to seconds:

$$\Delta JD_{0JDTIS} = 86400(JD_{TI} - JD_0).$$

(22) Compute the velocity increment due to trim phase:

$$\Delta V_{TRIMF} = |\mathbf{A}_{TRIM}|\Delta t_{TRIM}.$$

(23) Compute the estimate of mass expenditure on trim phase:

$$\Delta M_{TRIM} = M_0 \Delta V_{TRIMF}/V_E.$$

(24) Compute the mass estimate at final time of the trim phase:

$$M_{TRIMF} = M_0 - \Delta M_{TRIM}.$$

(25) Compute the thrust constraints on BP-CT-arc: $F_{THR} = 0.93 F_{max}$, $F_{BRF} = 0.57 F_{max}$.

(26) Compute the mass-flow rates:

$$\dot{M}_{THR} = F_{THR}/V_E, \quad \dot{M}_{BRF} = F_{BRF}/V_E.$$

(27) Compute the mass estimates:

$$M_{THR} = M_{TRIMF} - \dot{M}_{THR}\Delta t_{BRF}, \quad M_{BRF} = M_{THR} - \dot{M}_{BRF}\Delta t_{THR}.$$

(28) Compute $\mathbf{R}_{BRIP}$ and $\mathbf{V}_{BRIP}$ at t_I:

$$\mathbf{R}_{BRIP} = \left(r_p\cos(\theta_0),\ 0,\ \ -r_p\sin(\theta_0)\right),$$

$$\mathbf{V}_{BRIP} = \left(\sqrt{\frac{\mu}{r_p}}(1+e_{trans})\sin(\theta_0),\ 0,\ \sqrt{\frac{\mu}{r_p}}(1+e_{trans})\cos(\theta_0)\right).$$

Note that in the P63 ignition program these vectors are calculated directly in the inertial system by propagating the state forward from t_0 (about 10 min before the ignition) to t_I.
These vectors are computed at perilune of the entry orbit. It should be noted that the inertial coordinate system is not used in the BP-targeting procedures. It is convenient to choose the platform system to calculate the position and velocity vectors at the perilune at t_I. As described above, the platform coordinate system is defined at time t_0 and the $X_P Z_P$-plane of this system is the nominal trajectory plane. That's why the Y-components of all vectors defined in the platform system are zeros. The position and velocity vectors at the perigee of the entry orbit at t_I are computed as follows:

$$\mathbf{R}_{BRIP} = \left(r_p\cos(\theta_0),\ 0,\ \ -r_p\sin(\theta_0)\right),$$

$$\mathbf{V}_{BRIP} = \left(\sqrt{\frac{\mu}{r_p}}(1+e_{trans})\sin(\theta_0),\ 0,\ \sqrt{\frac{\mu}{r_p}}(1+e_{trans})\cos(\theta_0)\right).$$

(29) Save the position and velocity vectors computed at t_I:

$$\mathbf{R}_{BRIPX} = \mathbf{R}_{BRIP}, \quad \mathbf{V}_{BRIPX} = \mathbf{V}_{BRIP}.$$

(30) Compute the gravity vector in the platform system at t_I:

$$\mathbf{G}_{MP}(t_I) = -\mu_M/\left|\mathbf{R}_{BRIP}\right|^3\,\mathbf{R}_{BRIP},$$

(31) Propagate $\mathbf{R}_{BRIP}$, $\mathbf{V}_{BRIP}$ backwards from t_I to

$$t_0 = t_I - \Delta t_{BA} - \Delta t_{TRIM}\ sec.$$

No thrust is used in this propagation. Note that at t_0 the angle between the position vector and the LS vector should be about 45° to start the ignition algorithm.

(32) Propagate $\mathbf{R}_P(t_0)$, $\mathbf{V}_P(t_0)$ backwards from t_0 to t_x, and save the results, $\mathbf{R}_P(t_x)$, $\mathbf{V}_P(t_x)$:

$$t_0 - t_x = \pi\sqrt{\frac{a_{trans}^3}{\mu_M}} - \Delta t_{BA} - \Delta t_{TRIM}, \quad |\mathbf{R}_P(t_x)| = r_a,$$

$$|\mathbf{V}_P(t_x)| = \sqrt{\frac{\mu_M}{r_a}}.$$

(33) Set $\mathbf{T}_P^G = \mathbf{I}$ at t_I (for the first iteration).

(34) Set the Moon rotation rate vector in the local system: $\boldsymbol{\omega}_{ML} = (0,\ 0,\ \omega_M)$.

(35) Compute nominal landing site vector:

$$R_{LS} = R_M + H_L,$$

$$\begin{aligned} R_{XLSL} &= R_{LS}\cos\varphi_L\cos\theta_L, \\ R_{YLSL} &= R_{LS}\cos\varphi_L\sin\theta_L, \\ R_{ZLSL} &= R_{LS}\sin\varphi_L. \end{aligned}$$

$$\mathbf{R}_{LSL} = (R_{XLSL}, R_{YLSL}, R_{ZLSL}).$$

(36) Compute the time elapsed since July 1, 1969 in seconds at t_x:

$$t_{0SEC} = t_{0DAYS} * 86400.$$

(37) Compute the time elapsed since July 1, 1969 in seconds at t_{LAND}:

$$t_{CURRENT} = \Delta JD_{0JDLANDS}.$$

(38) Compute:

$$\mathbf{T}_I^L(t_{LAND}) = \mathbf{T}_I^L(t_{0SEC}, t_{CURRENT}).$$

(39) Compute the nominal LS vector in inertial system at t_{LAND}:

$$\mathbf{R}_{LSITLAND} = \left(\mathbf{T}_I^L(t_{LAND})\right)^T \mathbf{R}_{LSL}.$$

(40) Compute $\mathbf{T}_I^P$ at t_x:

$$\mathbf{T}_I^P = \mathbf{T}_I^P(\mathbf{R}_P(t_x), \mathbf{V}_P(t_x)).$$

(41) Compute the time elapsed since July 1, 1969 in seconds at t_I:

$$t_{CURRENT} = \Delta JD_{0JDTIS}.$$

(42) Compute $\mathbf{T}_I^L$ at t_I:

$$\mathbf{T}_I^L(t_I) = \mathbf{T}_I^L(t_{0SEC}, t_{CURRENT}).$$

(43) Compute $\mathbf{T}_L^P$ at t_I:

$$\mathbf{T}_L^P(t_I) = \mathbf{T}_I^P\left(\mathbf{T}_I^L(t_I)\right)^T.$$

(44) Compute the nominal LS vector, $\mathbf{R}_{LSP}$ at t_I:

$$\mathbf{R}_{LSP}(t_I) = \mathbf{T}_L^P(t_I)\mathbf{R}_{LSL}.$$

(45) Compute the Moon rotation rate vector in platform system at t_I:

$$\boldsymbol{\omega}_{MP}(t_I) = \mathbf{T}_L^P(t_I)\boldsymbol{\omega}_{ML}.$$

(46) Set the gravitational acceleration vector:

$$\mathbf{G_{MG}} = (G_M,\ 0,\ 0).$$

(47) Initialize the unit thrust vector at the BP-terminal point:

$$\mathbf{U}_{BRFG} = (\cos\theta_{PBRF},\ 0,\ -\sin\theta_{PBRF}),$$

where θ_{PBRF} is the angle between vertical X-axis of the platform system and thrust direction.

(48) First iteration set up for P63 terminal state $\mathbf{X}_{BRFG}$:

$$\begin{aligned}
\mathbf{R}_{BRFG} &= \mathbf{R}_{APIG}, \\
\mathbf{V}_{BRFG} &= \mathbf{V}_{APIG}, \\
\mathbf{A}_{BRFG} &= \frac{F_{BRF}}{M_{BRF}}\mathbf{U}_{BRFG} - \mathbf{G_{MG}}, \\
\mathbf{J}_{BRFG} &= \left(-(A_{XBRFG} + G_{MG})\frac{\dot{M}_{BRF}}{M_{BRF}},\ 0,\ -A_{ZBRFG}\frac{\dot{M}_{BRF}}{M_{BRF}}\right), \\
\mathbf{S}_{BRFG} &= (0,\ 0,\ 0).
\end{aligned}$$

In this work we propose different forms for the terminal jerk and snap vectors based on simple differentiation operation (see previous subsection). Note that $K_J = 1.2$ accounts for the X component of thrust [1]. The reason for using this specific value remains unknown to the authors.

(49) Set $\mathbf{T}_P^G(t_I) = \mathbf{I}$ (for the first iteration).
(50) Compute the target state $\mathbf{X}_{BRTG}$:

$$\mathbf{X}_{BRTG} = STM(0, TBRF)\mathbf{X}_{BRFG}.$$

(51) Initialize the unit thrust vector at t_I (for the first iteration):

$$\mathbf{U}_{BRIP} = (0,\ 0,\ -1).$$

(52) Copy the nominal landing site vector:

$$\mathbf{R}_{LSP1}(t_I) = \mathbf{R}_{LSP}(t_I).$$

(53) Set $T = -664$ s, $t = t_I$.
(54) Set: $N = 5$ (maximum number of iterations),

$$N_{GUIDCALLS} = 1,\ t_{CLOCK} = t_I,\ PHASE = 1.$$

(55) Correct the velocity $\mathbf{V}_{BRIP}$ for $\Delta t_{TRIM} = 26$ s of minimum thrust:

$$\mathbf{V}_{BRIP} \Leftarrow \mathbf{V}_{BRIP} + |\mathbf{A}_{TRIM}|\mathbf{U}_{BRCP}\Delta t_{TRIM}.$$

(56) Call P63-64 guidance algorithm to produce $\mathbf{A}_{BRCP}$, $\mathbf{U}_{BRCP}$, $\mathbf{W}_{BRCP}$ and update $\mathbf{T}_P^G$, $\mathbf{R}_{LSP}$ and T.
(57) Calculate: $N = N - 1$. If $N > 0$ go to the step, where $\mathbf{V}_{BRIP}$ is corrected for $\Delta t_{TRIM} = 26$ s of minimum thrust.
(58) Save:

$$\mathbf{R}_{BRIP},\quad \mathbf{V}_{BRIP},\quad \mathbf{G}_{BRIP},\quad T,\quad \mathbf{U}_{BRIP},$$
$$\omega_{MP}(t_I),\quad \mathbf{R}_{LSP}(t_I),\quad \mathbf{T}_P^G(t_I).$$

(59) Save mass and thrust acceleration:

$$M_I = M_{TRIMF},\quad A_{FCP} = F_{THR}/M_I.$$

(60) Set

$$N_{GUIDCALLS} = 1,\ t_{CLOCK} = t_I,\ PHASE = 1,\ QUARTIC = 0.$$

(61) Set the time step for integration of equations of motion: $\Delta t = 0.1\ s$.
(62) Propagate the state forwards from t_I of the braking phase with $\mathbf{R}_{BRIP}$ and $\mathbf{V}_{BRIP}$ to t' such that $F_{BR}(t') \leq 0.57F_{max}$, and the current time t is updated as $t \Leftarrow t + \Delta t$, where Δt is user defined quantity.

(a) Use P63-64 guidance algorithm with computed $\mathbf{X}_{BRTG}$ to find commanded $\mathbf{U}_{FCP}$, $\mathbf{T}_P^G$ and T on the interval $t \in [t_I, t']$. At initial time of the propagation one can set $T = T'\ sec$, and $F_{BR} = 0.93F_{max}$.

(b) Compute the magnitude of thrust, $F_{BR}(t)$ at a given time t:

$$M_{BR}(t) = M_{TRIMF} - \dot{M}_{THR}(t - t_I),$$

$$F_{BR}(t) = |\mathbf{A}_{BRCP}(t)| M_{BR}(t).$$

(c) The magnitude of thrust produced by P63-64 algorithm is greater than 100% of F_{max} for most of the time on braking phase and that's why the magnitude of the thrust acceleration commanded is not used until the following condition is satisfied: $F_{BR}(t) \leq 0.57F_{max}$.

(d) If $F_{BR} \leq 0.57F_{max}$ at $t = t'$, then save

$$THR' = T(t'), \quad TBRF' = THR' + \Delta t_{THR},$$

$$\mathbf{R}_G = \mathbf{R}_G(t'), \quad \mathbf{V}_G = \mathbf{V}_G(t'), \quad \mathbf{A}_G = \mathbf{A}_G(t'),$$

$$\mathbf{R}_P = \mathbf{R}_P(t'), \quad \mathbf{V}_P = \mathbf{V}_P(t'), \quad \mathbf{A}_P = \mathbf{A}_P(t'),$$

$T(t')$ is the output of P63-64 at the last call before the 120 s. flight from TCR to BRF.

(63) Compute the new jerk and snap vectors taking advantage of the fact that the BP-quartic starts from t' and is continued until t_{TBRT} (or THR' to 0 for 120 s.), where $t_{TBRT} = t' + |THR'|$ and $T = THR'$:

$$\begin{bmatrix} \mathbf{J}'_{BRTG} \\ \mathbf{S}'_{BRTG} \end{bmatrix} = \begin{bmatrix} -\frac{24}{T^3} & -\frac{18}{T^2} & -\frac{6}{T} & \frac{24}{T^3} & -\frac{6}{T^2} \\ \frac{72}{T^4} & \frac{48}{T^3} & \frac{12}{T^2} & -\frac{72}{T^4} & \frac{24}{T^3} \end{bmatrix} \times$$

$$\times \left[\mathbf{R}_{BRTG}\ \mathbf{V}_{BRTG}\ \mathbf{A}_{BRTG}\ \mathbf{R}_G\ \mathbf{V}_G\ \right]^T.$$

Note that in the description of Apollo Lunar-Descent Guidance these vectors are denoted as $\mathbf{J}_{BRTGA}$ and $\mathbf{S}_{BRTGA}$.

(64) Compute new $\mathbf{X}'_{BRFG}$:

$$\mathbf{X}'_{BRFG} = STM(TBRF', 0) \left[\mathbf{R}_{BRTG}\ \mathbf{V}_{BRTG}\ \mathbf{A}_{BRTG}\ \mathbf{J}'_{BRTG}\ \mathbf{S}'_{BRTG}\right]^T.$$

(65) Correct all estimates of the current phase parameters:

$$M'_{BRF} = M_{BRF} EXP \frac{|\mathbf{V}'_{BRFG} - \mathbf{V}_{BRFG}|}{V_E},$$

$$\mathbf{A}'_{BRFG} = \frac{F_{BRF}}{M'_{BRF}} \mathbf{U}_{BRFG} - \mathbf{G}_{MG},$$

$$\mathbf{J}'_{BRFG} = (J'_{XBRFG}, 0, -K_J\, A'_{ZBRFG} \frac{\dot{M}_{BRF}}{M'_{BRF}}),$$

$$\mathbf{S}'_{BRFG} = (S'_{XBRFG},\ 0,\ S'_{ZBRFG}),$$

where the vectors $\mathbf{V}_{BRFG}$ and $\mathbf{U}_{BRFG}$ are set up initially to given values.

(66) Check the state convergence using the following conditions:

$$Q_1 = \frac{|R'_{XBRFG} - R_{XBRFG}|}{|\mathbf{R}_{BRFG}|}, \quad Q_2 = \frac{|R'_{ZBRFG} - R_{ZBRFG}|}{|\mathbf{R}_{BRFG}|},$$

$$Q_3 = \frac{|V'_{XBRFG} - V_{XBRFG}|}{|\mathbf{V}_{BRFG}|}, \quad Q_4 = \frac{|V'_{ZBRFG} - V_{ZBRFG}|}{|\mathbf{V}_{BRFG}|},$$

$$Q_5 = \frac{|A'_{XBRFG} - A_{XBRFG}|}{|\mathbf{A}_{BRFG}|}, \quad Q_6 = \frac{|A'_{ZBRFG} - A_{ZBRFG}|}{|\mathbf{A}_{BRFG}|},$$

$$Q_7 = |THR - THR'|.$$

(67) Compute the sum of relative errors for position, velocity, acceleration and target referenced time:

$$S_{RVAT} = Q_1 + Q_2 + Q_3 + Q_4 + Q_5 + Q_6 + Q_7.$$

(68) If the condition $S_{RVAT} \leq \varepsilon_S$, with given $\varepsilon_S > 0$, is not satisfied, then go to the beginning of the iteration process (step 48) and repeat the process given above with new state vector computed in the previous iteration, $\mathbf{X}_{BRFG} = \mathbf{X}'_{BRFG}$. The iteration process is repeated until the latter conditions are satisfied. If these conditions are satisfied, then $\mathbf{R}_{BRIG}$, $\mathbf{V}_{BRIG}$, $\mathbf{X}_{BRFG}$ and $\mathbf{X}_{BRTG}$ are determined from the last iteration.

(69) Run AP-targeting with new

$$\mathbf{R}_{APIG} = \mathbf{R}_{BRFG}, \quad \mathbf{V}_{APIG} = \mathbf{V}_{BRFG}$$

to compute new (Fig. 6.2) $\mathbf{X}_{APTG}$.

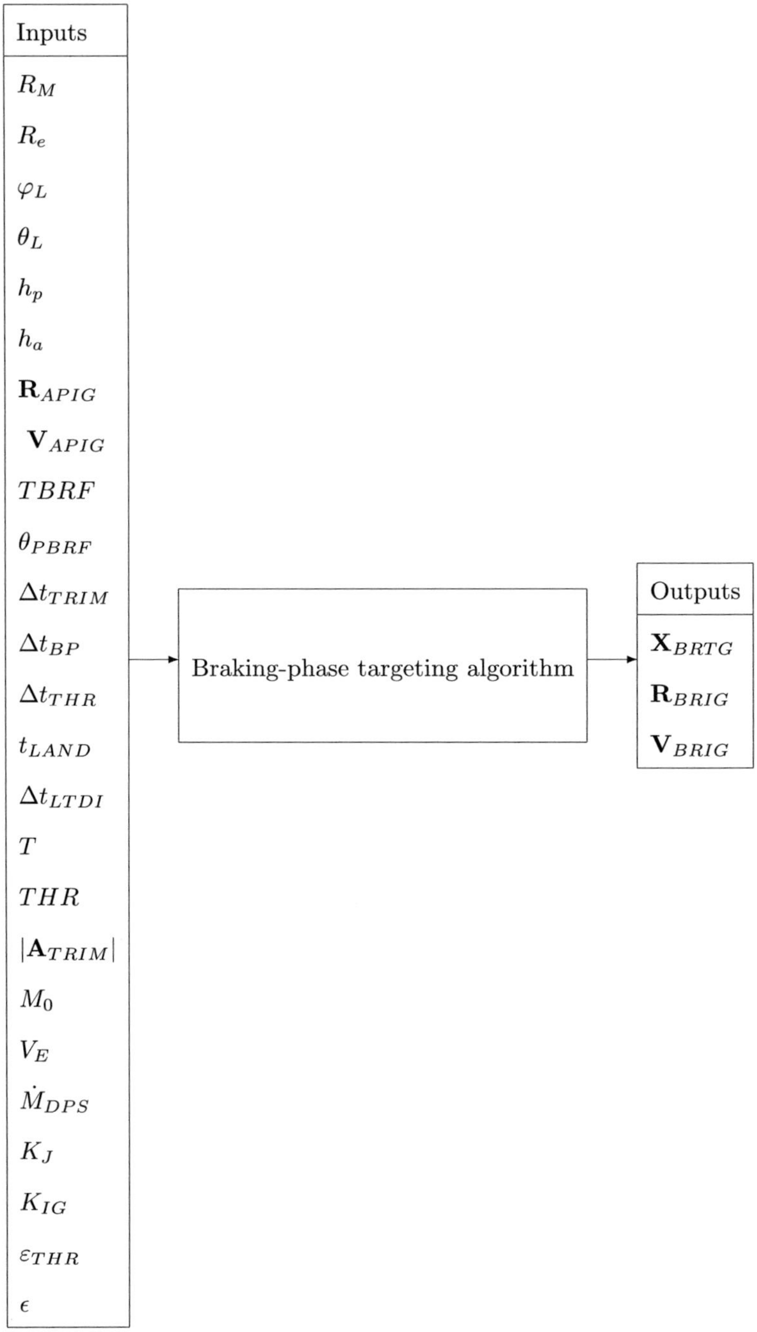

Fig. 6.2 Schematic illustration of the braking-phase targeting procedure

6.7 Alternative Targeting Procedure Using Backward Integration

The components of the complete terminal state, $\mathbf{X}_{BRFG}$ for the first iteration of P63 can be set up using the following estimates:

$$\begin{aligned}
\mathbf{R}_{BRFG} &= \mathbf{R}_{APIG}, \\
\mathbf{V}_{BRFG} &= \mathbf{V}_{APIG}, \\
\mathbf{A}_{BRFG} &= \frac{F_{BRF}}{M_{BRF}} \mathbf{U}_{BRFG} - \mathbf{G_{MG}}, \\
\mathbf{J}_{BRFG} &= (0, 0, -K_J \; A_{ZBRF} \frac{\dot{M}_{BRF}}{M_{BRF}}), \\
\mathbf{S}_{BRFG} &= (0,\ 0,\ 0),
\end{aligned} \tag{6.15}$$

where $\mathbf{G_{MG}} = (G_0,\ 0,\ 0)$.

Consequently, the complete target state $\mathbf{X}_{BRTG}$ is propagated forward from the terminal point using the quartic formulation and state transition matrix in the following form:

$$\mathbf{X}_{BRTG} = STM(0, TBRF)\mathbf{X}_{BRFG} \tag{6.16}$$

The backward propagation from the target point to the terminal point can be used to check if the corresponding part of the BP- quartic is being constructed properly using the matrix $STM(TBRF, 0)$. Similarly, the complete state at a throttle recovery instant can be determined using backward propagation from the terminal point:

$$\mathbf{X}_{BRHG} = STM(TTHROT, TBRF)\mathbf{X}_{BRFG}. \tag{6.17}$$

Once the terminal state of the braking phase is determined, we can integrate equations of motion (EQM) of LM backwards from the terminal instant t_{LBRF} to initial instant t_I on this phase. It should be noted that the integration starts from the terminal point because at this point and at the target point the state vector can be completely determined. So, for the first iteration, the portion of the BP-quartic from the terminal to target point is determined by setting up the state vector at the terminal point. As the BP constraints are not enough to form a closed-form solution, this state vector is considered as the initial state for the backward integration of (EQM).

First, compute the nominal landing site vector in the platform system at t_{LBRF}:

$$\mathbf{R}_{LSP} = (\mathbf{T}_P^L)^T \mathbf{R}_{LSL}, \tag{6.18}$$

where $\mathbf{T}_P^L = \mathbf{T}_P^L(t_{LBRF})$ is the transformation matrix from the platform (P) system to the local (L) coordinate system, $\mathbf{R}_{LSL}$ is the nominal landing site vector defined in the local system. As the integration of EQM is carried out in the platform system, it is necessary to transform the state vector from the guidance frame to the platform system. Consider position and velocity vectors, $\mathbf{R}_G$ and $\mathbf{V}_G$ defined at instant t in the guidance frame by following formulas:

$$\mathbf{R}_G = \mathbf{T}_P^G(\mathbf{R}_P - \mathbf{R}_{LSP}),$$

$$\mathbf{V}_G = \mathbf{T}_P^G(\mathbf{V}_P - \boldsymbol{\omega}_P \times \mathbf{R}_P),$$

where $\mathbf{T}_P^G = \mathbf{T}_P^G(t)$ is the transformation matrix from the platform system to the guidance coordinate system. By taking $t = t_{LBRF}$, these formulas can easily be solved for $\mathbf{R}_P$ and $\mathbf{V}_P$ in the form:

$$\mathbf{R}_P = (\mathbf{T}_P^G)^T \mathbf{R}_{RBRFG} + \mathbf{R}_{LSP},$$

$$\mathbf{V}_P = (\mathbf{T}_P^G)^T \mathbf{V}_{RBRFG} + \boldsymbol{\omega}_P \times \mathbf{R}_P, \tag{6.19}$$

where $\mathbf{T}_P^G = \mathbf{T}_P^G(t_{LBRF})$. These two vectors can be used as the initial position and velocity vectors for the backward integration from t_{LBRF} to t_I. The commanded unit thrust vector, $\mathbf{U}_{BRFG}$ can also be transformed similarly:

$$\mathbf{U}_{FCFP} = (\mathbf{T}_P^G)^T \mathbf{U}_{BRFG}. \tag{6.20}$$

The other important parameters for the integration of EQM are defined as follows:

$$\begin{aligned}
|\mathbf{A}_{FCFP}| &= \frac{F_{BRF}}{M_{BRF}},\\
t_{CLOCK} &= t_{LAND} - t_{LBRF},\\
t_{OLD} &= t_{CLOCK},\\
TTT &= TBRF,\\
t_0 &= t_{CLOCK},\\
t_{k+1} &= t_k + \Delta t,\\
M(t_0) &= M_{BRF}.
\end{aligned} \tag{6.21}$$

In particular, for better results, the time step Δt can be set to a certain value from the interval $[-1, -0.1]$. The integration of EQM, given as

$$\ddot{\mathbf{R}}_P = |\mathbf{A}_{FP}|\mathbf{U}_{FCP} + \mathbf{G}_{MP}. \tag{6.22}$$

It should be noted that in the integration process, the trajectory can be considered as quartic while $TTT \geq TTHROT$, where TTT is computed by the P63-P64 guidance algorithm (see section "Guidance procedures"). Let's denote the first value of TTT, obtained from the guidance algorithm and which does not satisfy this condition, by $THRA$, that is

$$THRA = TTT.$$

$THRA$ will be considered as an actual target referenced time, at which the throttle recovery occurs. At this time one can accept that the mass-flow rate and thrust magnitude are defined as

$$\dot{M} = \dot{M}_{THR}, \qquad F = F_{THR}.$$

When the inequality condition $TTT < TTHROT$ is selected, that is when the trajectory is not approximated as a quartic, these parameters are defined accordingly:

$$\dot{M} = \dot{M}_{BRF}, \qquad F = F_{BRF}.$$

Consequently, the mass and commanded thrust acceleration for the next cycle are computed using formulas:

$$M_{k+1} = M_k + |\dot{M}|Dt, \quad |\mathbf{A}_{FP}|_{k+1} = \frac{F}{M_{k+1}}. \tag{6.23}$$

where, according to Apollo guidance procedures, $\Delta t = 2$ s. This indicates that the guidance cycle is repeated every 2 s, while the integration of EQM is carried out every $\Delta t = 1$ s or less second. It means that the EQM are integrated with constant thrust acceleration for each 2 s interval, after which the guidance cycle provides with new matrix $T_P^G(t_{k+1})$, magnitude of the commanded thrust acceleration $|\mathbf{A}_{FCP}(t_{k+1})|$, commanded unit thrust and window vectors $\mathbf{U}_{FCP}(k+1)$, $\mathbf{U}_{WCP}(t_{k+1})$, and the target referenced time TTT assuming that at the last instant of the previous guidance cycle $t = t_k$. But during the braking phase, the EQM are integrated not with the use of commanded thrust acceleration. Instead, the integration uses actual thrust acceleration, computed by Eq. (6.23). The current simulation time, t_{SIM}, that is the time remaining until the landing is also updated every 2 s:

$$t_{SIM,k+1} = t_{SIM,k} + Dt.$$

At each step of the integration, the selenocentric altitude, H and gravity vector, $\mathbf{G}_P$ can be computed using formulas:

$$H = |\mathbf{R}_P| - R_e, \quad \mathbf{G}_P = -\frac{\mu_M}{|\mathbf{R}_P|^3}\mathbf{R}_P. \tag{6.24}$$

The integration is carried out until following conditions are satisfied:

$$H \simeq H_0, \qquad R_{ZP} \simeq R_{ZBRIP}. \tag{6.25}$$

If one or both of these conditions are not satisfied, then one can change t_I and/or $TTHROT$ and the integration starts over with these values. Other parameters remain the same and should not be changed in order to satisfy terminal conditions and other constraints, imposed on quartics. If these conditions are violated as a result of changes in t_I and/or $TTHROT$, then the new values for $TAPM$ and $TAPI$ are selected and the AP targeting is repeated and so on until the conditions Eq. (6.25) are satisfied. One may not exclude that the whole process of BP-targeting procedures is cumbersome and time consuming. So, when the conditions Eq. (6.25) are satisfied, the AP- and BP-targeting procedures are conducted together in order to provide continuity of the state vector at the point, where the corresponding quartics are connected. At an instant, when the conditions Eq. (6.25) are satisfied, one can determine the actual values of the position and velocity vectors, the commanded unit thrust vector, commanded and actual thrust acceleration vectors, mass, and target referenced time of the initial point of the braking phase:

$$\begin{aligned}
\mathbf{R}_{BRIPA} &= \mathbf{R}_P, \\
\mathbf{V}_{BRIPA} &= \mathbf{V}_P, \\
\mathbf{R}_{BRIGA} &= \mathbf{R}_G, \\
\mathbf{V}_{BRIGA} &= \mathbf{V}_G, \\
\mathbf{U}_{FCPBRI} &= \mathbf{U}_{FCP}, \\
|\mathbf{A}_{FCPBRI}| &= |\mathbf{A}_{FCP}|, \\
|\mathbf{A}_{FPBRI}| &= |\mathbf{A}_{THR}|, \\
M_{BRIA} &= M, \\
t_{ICLOCKA} &= t_{CLOCK}, \\
t_{SIMBRI} &= t_{SIM}, \\
TBRI &= TTT.
\end{aligned} \tag{6.26}$$

In the same manner, one can integrate EQM forwards with initial conditions given in Eq. (6.26) from the instant $t_{ICLOCKA}$ to $TBRF$. Note that when the target referenced time TTT satisfies the condition, $TTT \geq THRA$, the thrust is changed from F_{THR} to F_{BRF}. The integration is stopped as soon as the condition $TTT > TBRF$ is satisfied. At this point the achieved values of the main parameters are as follows:

$$\begin{aligned}
\mathbf{R}_{BRFPA} &= \mathbf{R}_P, \\
\mathbf{V}_{BRFPA} &= \mathbf{V}_P, \\
\mathbf{R}_{BRFGA} &= \mathbf{R}_G, \\
\mathbf{V}_{BRFGA} &= \mathbf{V}_G, \\
\mathbf{U}_{FCPBRFA} &= \mathbf{U}_{FCP}, \\
|\mathbf{A}_{FCPBRFA}| &= |\mathbf{A}_{FCP}|, \\
|\mathbf{A}_{FPBRFA}| &= |\mathbf{A}_{BRF}|, \\
M_{BRFA} &= M, \\
t_{APICLOCKA} &= t_{CLOCK}, \\
t_{SIMBRFA} &= t_{SIM}, \\
TBRFA &= TTT.
\end{aligned} \tag{6.27}$$

Note that the values of these parameters at $TBRFA$ may differ from those used at the beginning of the backward integration at target referenced time TBRF and consequently, the BP-quartic and AP-quartic parameters will obviously be different. Therefore, it is necessary to check if the constraints imposed on these quartics are exactly or closely satisfied. With this purpose we further construct new quartics based on the actual values obtained from the forward integration of EQM in the interval $[t_{ICLOCKA}, t_{APICLOCKA}]$, that is the values given in Eq. (6.27).

First, one can update the jerk and snap vectors at the target point of the braking phase using the position and velocity vectors obtained by the simulated trajectory, that is at an instant $t_{APICLOCKA}$:

$$\begin{bmatrix} \mathbf{J}_{TA} \\ \mathbf{S}_{TA} \end{bmatrix} = \begin{bmatrix} -\frac{24}{T^3} & -\frac{18}{T^2} & -\frac{6}{T} & \frac{24}{T^3} & -\frac{6}{T^2} \\ \frac{72}{T^4} & \frac{48}{T^3} & \frac{12}{T^2} & -\frac{72}{T^4} & \frac{24}{T^3} \end{bmatrix} \begin{bmatrix} \mathbf{R}_T \\ \mathbf{V}_T \\ \mathbf{A}_T \\ \mathbf{R}_{BRFGA} \\ \mathbf{V}_{BRFGA} \end{bmatrix}, \tag{6.28}$$

where $T = TBRFA$. This equation indicates that the achieved vectors $\mathbf{R}_{BRFGA}$, $\mathbf{V}_{BRFA}$ and the vectors $\mathbf{R}_T$, $\mathbf{V}_T$ and $\mathbf{A}_T$ defined at the target satisfy the same quartic formulation. The corresponding complete state at terminus achieved by the simulated trajectory can be found easily by following quartic propagation equation:

$$\mathbf{X}_{BRFGA} = \begin{bmatrix} \mathbf{R}_{BRFGA} \\ \mathbf{V}_{BRFGA} \\ \mathbf{A}_{BRFGA} \\ \mathbf{J}_{BRFGA} \\ \mathbf{S}_{BRFGA} \end{bmatrix} = STM(TBRF, 0) \begin{bmatrix} \mathbf{R}_T \\ \mathbf{V}_T \\ \mathbf{A}_T \\ \mathbf{S}_T \\ J_{XTA}, J_{YTA}, J_{ZT} \end{bmatrix}. \tag{6.29}$$

Then all estimates of the current phase parameters given in Eq. (6.15) can be corrected as follows:

$$\begin{aligned} M_{BRF} &= M_{MBRFA} EXP \frac{|\mathbf{V}_{BRFG}| - \mathbf{V}_G|}{V_E}, \\ |\mathbf{A}_{BRFG}| &= \frac{F_{BRF}}{M_{BRF}} \mathbf{U}_{BRFG} - \mathbf{G_M}, \\ J_{XBRFG} &= (J_{XBRFGA}, 0, -K_J\ A_{ZBRF} \frac{\dot{M}_{BRF}}{M_{BRF}}), \\ \mathbf{S}_{BRFG} &= (S_{XBRFGA},\ 0,\ S_{ZBRFGA}), \end{aligned} \tag{6.30}$$

where $\mathbf{G_M} = (G_0,\ 0,\ 0)$. After these corrections, one can check the state convergence using following conditions:

$$
\begin{aligned}
Q_1 &= \frac{|A_{ZBRFGA} - A_{ZBRFG}|}{|\mathbf{A}_{BRFG}|} \leq \varepsilon_{AZ}, \\
Q_2 &= \frac{|A_{XBRFGA} - A_{XBRFG}|}{|\mathbf{A}_{BRFG}|} \leq \varepsilon_{AX}, \\
Q_3 &= \frac{|V_{XBRFGA} - V_{XBRFG}|}{|\mathbf{V}_{BRFG}|} \leq \varepsilon_{VX}, \\
Q4 &= THRA - TTHROT \leq \varepsilon_{THR},
\end{aligned}
\tag{6.31}
$$

where ε_{AX}, ε_{AZ}, ε_{VX}, ε_{THR} are given parameters. If any of these conditions is not satisfied, then the next iteration is started with new state:

$$
\mathbf{X}_{BRFG} = \mathbf{X}_{BRFGA}. \tag{6.32}
$$

The iteration process is repeated again until the conditions Eq. (6.31) are satisfied. If these conditions are satisfied, then the BP initial, terminal and target states are determined from the last iteration.

After this BP-targeting procedure, it is necessary to update the AP targets as the initial conditions for the AP quartic have changed and consequently, the corresponding terminal and target states will, in general, be different from those considered for the first iteration. The new initial state for this phase is defined as

$$
\mathbf{X}_{APIG} = \mathbf{X}_{BRFG}. \tag{6.33}
$$

References

1. Klumpp AR (1971) "Apollo Lunar-Descent Guidance," Apollo Guidance Navigation Control, R-695. Massachusetts Institute of Technology, MIT Instrumentation Laboratory Report, Cambridge, MA
2. Apollo 12 Mission Report. Trajectory Reconstruction and Post-flight Analysis, vol 1, 113 p, NASA-TM-X-70272, Houston, Texas (1970)

Chapter 7
P63 Ignition Algorithm

7.1 Preliminary Comments

The main on-board algorithms of the Apollo landers were P63 ignition, P63-64 guidance, P66 horizontal guidance and P66 vertical guidance, throttle and attitude control algorithms. Among them, the P63 ignition and P63-64 guidance algorithms are of a critical importance to the proper understanding and execution of the targeting algorithms. Similar to the BP-targeting algorithm, the ignition algorithm involves the explicit computations of the transformation matrices and the central angle between the lander's position vector and the LS vector. These computations in turn lead to a modification of the original algorithm, and add the corrected matrix of transformation between the platform and guidance systems to the list of the output parameters. However, the list of the main procedures of the on-board guidance algorithm remain unchanged. In this section, the modified version of the ignition algorithm with its inputs and outputs, and a brief description of the guidance algorithm in the context of the previous sections are presented.

7.2 Utility of the Ignition and Guidance Algorithms

The ignition algorithm is executed 10 min before the ignition at t_0, which is computed in the BP- targeting algorithm, but with $t_x \neq 0$:

$$t_0 = t_x + \pi\sqrt{\frac{a_{trans}^3}{\mu_M}} - \Delta t_{BA} - \Delta t_{TRIM}, \tag{7.1}$$

where t_x is the ground elapsed time at apolune of the entry orbit (measured relative to lift-off time). The algorithm starts with the pre-ignition algorithm, which determines the time to start the ignition algorithm. The purpose of this algorithm is to compute

© The Author(s), under exclusive license to Springer Nature Switzerland AG 2026

D. M. Azimov and R. H. Bishop, *Apollo-Class Targeting and Guidance for Lunar Powered Descent and Precision Landing*, Springer Aerospace Technology,
https://doi.org/10.1007/978-3-031-91088-3_7

the corrected unit thrust command to be used by P63-64 guidance algorithm at the first guidance pass, accurate clock-times of the first guidance pass, the ignition and the ullage. This algorithm also computes the position, velocity and gravity vectors in the platform system, and the corrected transformation matrix from the platform coordinate system to the guidance coordinate system.

The P63-64 guidance algorithm starts at time t_I which will be computed in the preceding P63 ignition algorithm and its execution is terminated when the lander reaches the AP-terminal point at $TAPF$. Basically P63 and P64 algorithms are run during the braking and approach phases respectively. The purpose of the P63-64 algorithm is to provide the unit thrust and unit window vectors, thrust acceleration magnitude, redesignation of the landing site, and the orientation of the platform coordinate system with respect to the guidance coordinate system. Note that there still exist some minor and unclear aspects of the original Apollo algorithms mostly associated with the specific values of certain parameters [1].

7.3 Modified Ignition Algorithm, Inputs and Outputs

7.3.1 Ignition Algorithm Inputs and Outputs

The P63 ignition algorithm inputs and outputs are given in Tables 7.1 and 7.2.

The flow-chart of the P63 ignition algorithm is provided in Fig. 7.1. The detailed and itemized description of the algorithm is given in the next section.

Table 7.1 Inputs of P63 Ignition Algorithm of Apollo 11

Current clock time tag	t
Transformation matrix $\mathbf{T}_L^I$ at t	$\mathbf{T}_L^I(t)$
Initial angle between the axes X_I and X_L at t_I	α_0
Current state on coasting trajectory	$\mathbf{R}_I(t),\ \mathbf{V}_I(t)$
Estimated clock-time of the first guidance pass	$t_I = 633.5$ s
Target referenced time at the first guidance pass	$T = -664$ s
Duration of ullage	$\Delta t_{ULLAGE} = 7.5$ s
LS radius-vector's magnitude, latitude, longitude	$R_e,\ \varphi_L,\ \theta_L$
Terminal target referenced time	$TBRF = -60.0$ s
Magnitude of thrust acceleration on trim phase	$\lvert\mathbf{A}_{TRIM}\rvert = 0.35$ m/s^2

Table 7.2 Outputs of P63 Ignition Algorithm of Apollo 11

Ignition and ullage times	$t_I,\ t_U$
Position, velocity and gravity vectors	$\mathbf{R}_P(t_I), \mathbf{V}_P(t_I),\ \mathbf{G}_{MP}$
Transformation matrices	$\mathbf{T}_L^I(t_I),\ \mathbf{T}_I^P(t_I),\ \mathbf{T}_P^L(t)$
Corrected transformation matrix	$\mathbf{T}_P^G(t_I)$
Corrected unit thrust vector at ignition	$\mathbf{U}_{BRIP}$

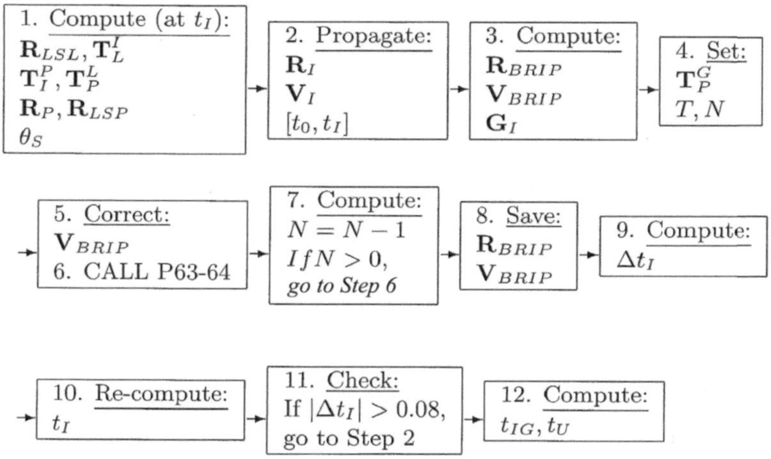

Fig. 7.1 Flow-chart of the P63 ignition algorithm

7.3.2 P63 On-Board Ignition Algorithm

(1) Pre-ignition algorithm.
Compute the angle $\theta_s(t)$ between $\mathbf{R}_P(t)$ and $\mathbf{R}_{LSP}(t)$ at current time $t = t_0$:

(a) Compute the landing site vector in the local system:
$\mathbf{R}_{LSL} = (R_{XLSL}, R_{YLSL}, R_{ZLSL})$

$$R_{XLSL} = R_e \cos\theta_L \cos\varphi_L,$$

$$R_{YLSL} = R_e \cos\theta_L \sin\varphi_L,$$

$$R_{ZLSL} = R_e \sin\theta_L.$$

(b) Compute $\mathbf{T}_L^I(t),\ \mathbf{T}_I^P(t),\ \mathbf{T}_P^L(t)$.

(c) Compute:
$\mathbf{R}_P(t) = \mathbf{T}_I^P(t)\mathbf{R}_I(t), \quad \mathbf{R}_{LSP}(t) = \left(\mathbf{T}_P^L(t)\right)^T \mathbf{R}_{LSL}.$

(d) Compute the angle $\theta_s(t)$ between $\mathbf{R}_P(t)$ and $\mathbf{R}_{LSP}(t)$ at t:

$$\cos\theta_s(t) = \frac{\mathbf{R}_P(t)\mathbf{R}_{LSP}(t)}{|\mathbf{R}_P(t)|\,|\mathbf{R}_{LSP}(t)|}.$$

(e) Check if $\theta_s(t) \le 45°$. If "No", then go to the first step and follow the procedure given above for next time step, which can be selected arbitrarily by the crew or can be set to 1 sec. The pre-ignition algorithm is completed here.

(2) Compute $\mathbf{T}_L^I(t_I)$, $\mathbf{T}_I^P(t_I)$, $\mathbf{T}_P^L(t_)$.
(3) Compute the landing site vector $\mathbf{R}_{LSP}$ at t_I:

$$\mathbf{R}_{LSP}(t_I) = \mathbf{T}_L^P(t_I)\mathbf{R}_{LSL}.$$

(4) Compute $\mathbf{R}_{BRII}$ and $\mathbf{V}_{BRII}$ by forward propagation from the current time $t = t_0$ to t_I.
(5) Compute $\mathbf{R}_{BRIP}$ and $\mathbf{V}_{BRIP}$ at t_I:
Option 1:

$$\mathbf{R}_{BRIP}(t_I) = \mathbf{T}_I^P(t_I)\mathbf{R}_{BRII}, \quad \mathbf{V}_{BRIP}(t_I) = \mathbf{T}_I^P(t_I)\mathbf{V}_{BRII}.$$

Option 2:
Propagate the state backwards from current $t = t_I$ to $t_0 = t_I - 10.0$ min to find the position and velocity vectors of LM when P63 starts. The current given position and velocity correspond to the perilune of the entry orbit which is assumed to be $h_p = 15$ km (perilune) by $h_a = 120$ km (apolune) above the surface of the Moon. It is also assumed that the central angle between the perilune and the nominal landing site is approximately $\theta_0 = 16$ degrees. So the initial conditions to start the backward propagation are as follows:

$$\mathbf{R}_{BRIP}(t_I) = \left((R_e + h_p)\sin(\theta_0),\ 0,\ (R_e + h_p)\cos(\theta_0)\right),$$

$$\mathbf{V}_{BRIP}(t_I) = \left(\sqrt{\frac{\mu}{R_e + h_p}}\cos(\theta_0),\ 0,\ -\sqrt{\frac{\mu}{R_e + h_p}}\sin(\theta_0)\right).$$

(6) Compute the gravity vector in the inertial and platform coordinate systems:

$$\mathbf{G}_{MI}(t_I) = -\mu/|\mathbf{R}_{BRII}|^3\,\mathbf{R}_{BRII}, \quad \mathbf{G}_{MP}(t_I) = \mathbf{T}_I^P(t_I)\mathbf{G}_{MI}(t_I).$$

(7) Set $\mathbf{T}_P^G = [\mathbf{I}]$, $\mathbf{U}_{BRIP} = (0,\ 0,\ -1)$.
(8) Set $T = -664$ s, $t = t_0$, and $N = 5$.
(9) Correct $\mathbf{V}_{BRIP}$ for $\Delta t_{TRIM} = 26$ s. of min. thrust:

$$\mathbf{V}_{BRIP} \Leftarrow \mathbf{V}_{BRIP} + |\mathbf{A}_{TRIM}|\mathbf{U}_{BRICP}\Delta t_{TRIM}.$$

(10) Call P63-64 guidance algorithm.
(11) Calculate: $N = N - 1$. If $N > 0$ go to step 8.
(12) Save: $\mathbf{R}_{BRIP}$, $\mathbf{V}_{BRIP}$.
(13) Compute correction to t_I using trajectory dispersions:

$$\Delta t_I = K_x \left(R_{XG}(t_I) - R_{XBRIG}\right) +$$

$$K_y R_{YG}^2(t_I) + \left(R_{ZG}(t_I) - R_{ZBRIG}\right) +$$

$$K_v \left(|V_G(t_I)| - |V_{BRIG}|\right) + \left(V_{ZG}(t_I) - K_x V_{XG}(t_I)\right),$$

where R_{XBRIG}, R_{ZBRIG}, $|\mathbf{V}_{BRIG}|$ are the outputs of the BP- targeting algorithm. Here one can accept that $K_x = K_y = K_v = 1$. The original values of these coefficients are remaining unknown to the authors.
(14) Re-compute t_I: $t_I \Leftarrow t_I + \Delta t_I$.
(15) Check the condition: if $|\Delta t_I| > 0.08$ s, then go to step 2.
(16) Compute necessary clock-times for the ignition and the ullage:
$t_{IG} = t_I - \Delta t_{TRIM}$ and $t_U = t_{IG} - \Delta t_{ULLAGE}$,
where $\Delta t_{ULLAGE} = 7.5$ s.

Reference

1. "Guidance System Operations Plan For Manned LM Earth Orbital and Lunar Missions Using Program Luminary 1C (LM131 Rev. 1)," Section 5. Guidance Equations. (Rev.8). Apollo Guidance Navigation Control, R-567, MIT, Cambridge, MA, April 1970

Chapter 8
P63-64 Powered Descent Guidance

8.1 Comments

In this chapter, the guidance algorithm is described in the context of the previous sections with necessary details and explanations. For the original description of this algorithm, the reader is referred to Klumpp's work in Ref. [1]. As mentioned in the previous chapters, the execution time for the P63-64 guidance algorithm is computed in the ignition algorithm. The guidance cycle on BP is repeated every $Dt = 2\,\text{s}$ and on AP—every second, while orbital integration is carried out every 1 or less second. It is necessary to distinguish the guidance cycle which is used in the targeting process on the ground and on-board. The results of the guidance cycle in the ground procedures are used iteratively to determine initial, terminal and target states of the braking phase. The on-board guidance algorithm produces the re-designated landing site, the commanded acceleration, the commanded unit thrust and unit window vectors, the updated target referenced time, and the updated transformation matrix, $\mathbf{T}_P^G$. Structurally, both guidance routines are the same, but the ground guidance cycle is used repeatedly for the same phase in the iteration process, whereas the on-board guidance routine is used only once for each cycle as the maneuver progresses. The latter algorithm is executed by an on-board computer starting at t_I (the clock-time of the first guidance pass, $t_{CLOCK} = t_I$) with $TTT = TBRI$. A significant portion of the nominal landing trajectory can be determined by employing a quartic polynomial formulation. This formulation satisfies five constraints on the nominal trajectory. This trajectory can be constructed backwards from the target point. The vectors $\mathbf{R}_G$, $\mathbf{V}_G$, $\mathbf{A}_G$ on the nominal trajectory can be determined in the guidance system using the quartic formulation.

© The Author(s), under exclusive license to Springer Nature Switzerland AG 2026

D. M. Azimov and R. H. Bishop, *Apollo-Class Targeting and Guidance for Lunar Powered Descent and Precision Landing*, Springer Aerospace Technology,
https://doi.org/10.1007/978-3-031-91088-3_8

8.2 Guidance Equations

8.2.1 Commanded Thrust Acceleration

As mentioned in the previous section, the quartic polynomial formulation allows us to satisfy five constraints on the reference trajectory, which can be propagated backwards from a target point. One can determine the position, velocity and acceleration vectors, $\mathbf{R}_G$, $\mathbf{V}_G$ and $\mathbf{A}_G$ on the reference trajectory in the following quartic polynomial formulation:

$$\begin{aligned}
\mathbf{R}_G &= \mathbf{R}_{TG} + T\mathbf{V}_{TG} + \frac{T^2}{2}\mathbf{A}_{TG} + \frac{T^3}{6}\mathbf{J}_{TG} + \frac{T^4}{24}\mathbf{S}_{TG}, \\
\mathbf{V}_G &= \mathbf{V}_{TG} + T\mathbf{A}_{TG} + \frac{T^2}{2}\mathbf{J}_{TG} + \frac{T^3}{6}\mathbf{S}_{TG}, \\
\mathbf{A}_G &= \mathbf{A}_{TG} + T\mathbf{J}_{TG} + \frac{T^2}{2}\mathbf{S}_{TG},
\end{aligned} \tag{8.1}$$

where T is the target referenced time, the vectors $\mathbf{R}_{TG}$, $\mathbf{V}_{TG}$, $\mathbf{A}_{TG}$, $\mathbf{J}_{TG}$, $\mathbf{S}_{TG}$ are position, velocity, acceleration, jerk and snap vectors defined at the target point. Note, that $T = 0$ at the target point, which means that T is negative during the maneuver.

Once the vectors $\mathbf{R}_G$, $\mathbf{V}_G$ and $\mathbf{A}_G$ are defined as in Eq. (8.1), then one can define the acceleration to be commanded as follows. The acceleration to be commanded at any point consists of the acceleration on the reference trajectory, two feedback terms, proportional to position and velocity deviations from the reference trajectory:

$$\begin{aligned}
\mathbf{A}_{CG} = {} & \mathbf{A}_{TG} + T\mathbf{J}_{TG} + \frac{T^2}{2}\mathbf{S}_{TG} - \\
& (\mathbf{V}_G - \mathbf{V}_{TG} - T\mathbf{A}_{TG} - \frac{T^2}{2}\mathbf{J}_{TG} - \frac{T^3}{6}\mathbf{S}_{TG})\frac{K_V}{T} - \\
& (\mathbf{R}_G - \mathbf{R}_{TG} - T\mathbf{V}_{TG} - \frac{T^2}{2}\mathbf{A}_{TG} - \frac{T^3}{6}\mathbf{J}_{TG} - \frac{T^4}{24}\mathbf{S}_{TG})\frac{K_R}{T^2},
\end{aligned} \tag{8.2}$$

where K_R and K_V are the feedback gains. Equation (8.2) is the implicit guidance equation. The reference trajectory corresponds to a quartic, but the trajectory, generated by Eq. (8.2) is not a quartic. This equation can be rewritten as a quartic and made as the explicit guidance equation by specifying the feedback gains, K_R and K_V. By regrouping the similar terms, one can rewrite Eq. (8.2) in the form:

$$
\begin{aligned}
\mathbf{A}_{CG} = & -\frac{K_R}{T^2}\mathbf{R}_G - \frac{K_V}{T}\mathbf{V}_G + \\
& \frac{K_R}{T^2}\mathbf{R}_{TG} + \frac{K_R + K_V}{T}\mathbf{V}_{TG} + \left(1 + K_V + \frac{K_R}{2}\right)\mathbf{A}_{TG} + \\
& \left(1 + \frac{K_V}{2} + \frac{K_R}{6}\right)T\mathbf{J}_{TG} + \left(\frac{1}{2} + \frac{K_V}{6} + \frac{K_R}{24}\right)T^2\mathbf{S}_{TG}.
\end{aligned} \tag{8.3}
$$

The gains K_R and K_V can be selected by equating the coefficients of $\mathbf{J}_{TG}$ and $\mathbf{S}_{TG}$ in Eq. (8.3) to zero:

$$1 + \frac{K_V}{2} + \frac{K_R}{6} = 0,$$

$$\frac{1}{2} + \frac{K_V}{6} + \frac{K_R}{24} = 0.$$

By solving these equations for K_R and K_V, it can easily be obtained that

$$K_R = 12, \qquad K_V = -6.$$

Consequently, Eq. (8.3) is rewritten as

$$\mathbf{A}_{CG} = \frac{12}{T^2}(\mathbf{R}_{TG} - \mathbf{R}_G) + \frac{6}{T}(\mathbf{V}_{TG} - \mathbf{V}_G) + \mathbf{A}_{TG}. \tag{8.4}$$

Let $T_P = T + t_d$, where t_d is transport delay, T_P is the predicted time. The acceleration of the quartic at the predicted time T_P is the acceleration to command at the current time T in order to realize the quartic profile. The constraint for the actual trajectory to be a quartic can be given as

$$\begin{bmatrix} \mathbf{R}_G \\ \mathbf{V}_G \end{bmatrix} = \begin{bmatrix} 1 & T & \frac{T^2}{2} & \frac{T^3}{6} & \frac{T^4}{24} \\ 0 & 1 & T & \frac{T^2}{2} & \frac{T^3}{6} \end{bmatrix} \begin{bmatrix} \mathbf{R}_{TG} \\ \mathbf{V}_{TG} \\ \mathbf{A}_{TG} \\ \mathbf{J}_{TA} \\ \mathbf{S}_{TA} \end{bmatrix},$$

where $\mathbf{R}_G$ and $\mathbf{V}_G$ are the current position and velocity vectors on the actual trajectory, and the jerk and snap vectors, $\mathbf{J}_{TA}$ and $\mathbf{S}_{TA}$ are considered as those which would be achieved at the target point on the actual trajectory. These two vectors can be found from the latter equation in the form:

$$\begin{bmatrix} \mathbf{J}_{TA} \\ \mathbf{S}_{TA} \end{bmatrix} = \begin{bmatrix} -\frac{24}{T^3} & -\frac{18}{T^2} & -\frac{6}{T} & \frac{24}{T^3} & -\frac{6}{T^2} \\ \frac{72}{T^4} & \frac{48}{T^3} & \frac{12}{T^2} & -\frac{72}{T^4} & \frac{24}{T^3} \end{bmatrix} \begin{bmatrix} \mathbf{R}_{TG} \\ \mathbf{V}_{TG} \\ \mathbf{A}_{TG} \\ \mathbf{R}_G \\ \mathbf{V}_G \end{bmatrix}. \tag{8.5}$$

So, the acceleration to be commanded at the current time T and realized at the predicted time T_P is given as

$$\mathbf{A}_{CG} = \mathbf{A}_{TG} + T_P \mathbf{J}_{TGA} + \frac{T_P^2}{2}\mathbf{S}_{TGA}. \tag{8.6}$$

By substituting Eq. (8.5) into Eq. (8.6), one can find the thrust acceleration to be commanded at the current time T and realized at the predicted time T_P

$$\mathbf{A}_{CG} = 12\frac{(3p^2 - 2p)}{T^2}(\mathbf{R}_{TG} - \mathbf{R}_G) + 6\frac{(4p^2 - 3p)}{T}\mathbf{V}_{TG} +$$

$$6\frac{(2p^2 - p)}{T}\mathbf{V}_G + (6p^2 - 6p + 1)\mathbf{A}_{TG}, \tag{8.7}$$

where $\mathbf{A}_{CG}$ is the commanded thrust acceleration, $\mathbf{R}_{TG}, \mathbf{V}_{TG}, \mathbf{A}_{TG}$ are the target position, velocity and acceleration respectively, $\mathbf{R}_G$ and $\mathbf{V}_G$ are the current position and velocity vectors, and $p = \frac{T_P}{T}$. Equation (8.7) is the main P63 and P64 Lunar-descent guidance formula for the commanded thrust acceleration [1].

8.2.2 *Computation of Target Referenced Time*

It should be noted that in Eqs. (8.4) and (8.7), the T is not constrained. This means that T is arbitrary and can be computed to satisfy an additional boundary constraint. In Apollo guidance, this constraint is obtained from the first equation of Eq. (8.5) written for the Z-components of the vectors in the following form:

$$T^3 J_{ZTG} + 6T^2 A_{ZTG} + T(18V_{ZTG} + 6V_{ZG}) + 24(R_{ZTG} - R_{ZG}) = 0.$$

One negative root of this equation is used as the desired value of T.

8.2.3 *Computation of TPG_Y*

In this subsection we derive the the vector TPG_Y, which consists of the elements of the second row of $\mathbf{T}_P^G$. Note that $\mathbf{T}_P^G$ is calculated in the P63-P64 guidance algorithm described below. If $\mathbf{X}_F$ is the state vector defined at the terminal point in a non-inertial Moon fixed rotating coordinate system, then the current position and velocity vectors, $\mathbf{R}$ and $\mathbf{V}$ at terminus referenced time T in the same coordinate system can be defined as

$$\mathbf{R} = \mathbf{R}_F + \mathbf{V}_F T + \mathbf{A}_F \frac{T^2}{2} + \mathbf{J}_F \frac{T^3}{6} + \mathbf{S}_F \frac{T^4}{24},$$

$$\mathbf{V} = \mathbf{V}_F + \mathbf{A}_F T + \mathbf{J}_F \frac{T^2}{2} + \mathbf{S}_F \frac{T^3}{6}.$$

It can be shown that the jerk vector can be found from these expressions as

$$\mathbf{J}_F = \frac{24}{T^3}\left(\mathbf{R} - \mathbf{V}\frac{T}{4} - \mathbf{R}_F - \frac{3}{4}\mathbf{V}_F T - \frac{1}{2}\mathbf{A}_F \frac{T^2}{2}\right).$$

As

$$R_{XF} = V_{XF} = A_{XF} = J_{XF} = 0,$$

then

$$\left(\mathbf{R} - \mathbf{V}\frac{T}{4}\right)_Y = 0,$$

Consider the guidance system as a non-inertial rotating coordinate system. As the current position and velocity vectors, **R** and **V** are defined in this system, then one can show that

$$\mathbf{R}_G = \mathbf{T}_P^G \left(\mathbf{R}_P - \mathbf{R}_{LSP}\right),$$

$$\mathbf{V}_G = \mathbf{T}_P^G \left(\mathbf{V}_P - \omega_M \mathbf{R}_{LSP}\right).$$

In this case the cross product

$$\mathbf{R}_{LSP} \times \left(\mathbf{R} - \mathbf{V}\frac{T}{4}\right) = \mathbf{R}_{LSP} \times \mathbf{T}_P^G \left[\left(\mathbf{R}_P - \mathbf{R}_{LSP}\right) - \left(\mathbf{V}_P - \omega_M \mathbf{R}_{LSP}\right)\frac{T}{4}\right]$$

can be used to form the matrix $\mathbf{T}_P^G$ with $T = TTT$. Consequently,

$$TPG_Y(t) =$$

$$= \mathbf{unit}\left[\mathbf{R}_{LSP}(t) \times [\mathbf{R}_P(t) - \mathbf{R}_{LSP}(t) - K\left(\mathbf{V}_P(t) - \omega_{MP}(t) \times \mathbf{R}_P(t)\right)\frac{TTT}{4}\right].$$

8.3 P63-64 Guidance Algorithm, Inputs and Outputs

8.3.1 Algorithm Inputs and Outputs

The algorithm inputs and outputs are given in Tables 8.1 and 8.2.

Table 8.1 The Apollo 11 guidance algorithm inputs

Azimuth and elevation command counts	$N_{CAZ},\ N_{CEL}$
Current position and velocity vectors	$\mathbf{R}_P,\ \mathbf{V}_P$
Gravity and LS vectors	$\mathbf{G}_{MP},\ \mathbf{R}_{LSP}$
Clock time tag of current state	t
Target referenced time	$TBRI = -664$ s
Transformation matrices	$\mathbf{T}_L^I,\ \mathbf{T}_I^P,\ \mathbf{T}_P^L,\ \mathbf{T}_P^G$

Table 8.2 The Apollo 11 guidance algorithm outputs

Thrust acceleration command	A_{FCP}
Unit thrust and window commands	$\mathbf{U}_{FCP},\ \mathbf{U}_{WCP}$
Updated target referenced time	T
Updated transformation matrix	$\mathbf{T}_P^G$
Re-designated LS vector	$\mathbf{R}_{LSP}$

8.3.2 Guidance Algorithm

1. ***Clock time and Moon rotational velocity***
   The procedure starts with determination of the old clock-time and transformation of the Moon rotational velocity from the local frame to the platform frame:

$$t = t_{CLOCK},$$

$$\omega_{ML} = (0,\ 0,\ \omega_M),$$

$$\omega_{MP} = (\mathbf{T}_P^L)^T \omega_{ML}.$$

$$\mathbf{T}_L^P(t) = \mathbf{T}_I^P(t)\mathbf{T}_L^I(t).$$

2. ***Position and velocity vectors at current time***
   These vectors are computed based on the current position and velocity vectors defined in the inertial coordinate system:

$$\mathbf{R}_P(t) = \mathbf{T}_I^P \mathbf{R}_I(t),$$

$$\mathbf{V}_P(t) = \mathbf{T}_I^P \mathbf{V}_I(t).$$

3. ***Landing site vector update***
   The landing site vector in the platform system is updated as follows:

$$\mathbf{R}_{LSP}(t) = |\mathbf{R}_{LSP}(t)|\mathbf{unit}[\mathbf{R}_{LSP}(t) + (t - t_{OLD})\omega_{MP} \times \mathbf{R}_{LSP}(t)],$$

where t_{OLD} is the clock-time of the previous cycle (see step 11 below). Note, that the phrase **unit** is used to simplify the notation and determined as the operation of computing the unit vector of a vector given after this phrase.

4. ***Line of site vector***
   The line of site vector, $\mathbf{R}_{LOSP}$ at current time, t is computed as follows:

$$\mathbf{R}_{LOSP} = \textbf{unit}\left(\mathbf{R}_{LSP}(t) - \mathbf{R}_P(t)\right),$$

$$(update)\quad \mathbf{R}_{LOSP} =$$

$$= \mathbf{R}_{LOSP} + 0.01745NCAZ\mathbf{T}_{PY}^{B} + 0.01745NCEL\left(\mathbf{T}_{PY}^{B} \times \mathbf{R}_{LOSP}\right),$$

$$NCAZ = 0,$$

$$NCEL = 0,$$

$$R_{XLOSP} = min(R_{XLOSP} - 0.02),$$

$$\mathbf{R}_{LOSP} = \textbf{unit}(\mathbf{R}_{LOSP}),$$

If the commander manipulates the controller in the automatic mode, then the azimuth command count, $NCAZ$ and the elevation command count, $NCEL$ are incremented or decremented according to the direction of manipulation. If these quantities are zero, then $\mathbf{R}_{LOSP}$ (updated) will be the same for subsequent cycles and the redesignation algorithm has no effect [1]. This case can be considered for simplicity of the simulation.

5. ***Transformation matrix from platform to body system***
   This transformation matrix is formed based on the following unit vectors:

$$\begin{aligned} \mathbf{u}_{XB} &= \mathbf{U}_{FCP}, \\ \mathbf{u}_{YB} &= \textbf{unit}\left(\mathbf{U}_{WCP} \times \mathbf{u}_{XB}\right), \\ \mathbf{u}_{ZB} &= \mathbf{u}_{XB} \times \mathbf{u}_{YB}. \end{aligned}$$

For $\mathbf{U}_{FCP}$ see Step 7 of the P63 ignition algorithm and Steps 13 and 14 of this algorithm. Then

$$\mathbf{T}_P^B = \begin{bmatrix} \mathbf{u}_{XB} \\ \mathbf{u}_{YB} \\ \mathbf{u}_{ZB} \end{bmatrix}.$$

Here, the commanded window vector, $\mathbf{U}_{WCP}$ is computed as follows (see Fig. 3.3 of Chap. 3):

$$\mathbf{u}_{XB} = \mathbf{U}_{FCP},$$

$$\mathbf{F}_{RP} = \textbf{unit}\left(\mathbf{u}_{XB} \times \mathbf{T}_{PY}^{G}\right),$$

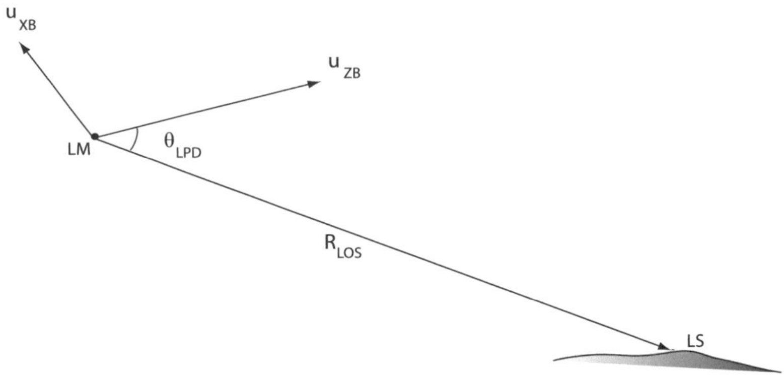

Fig. 8.1 Illustration of the landing point re-designator angle

$$PROJ = (\mathbf{R}_{LOSP} \times \mathbf{u}_{XB})\,\mathbf{T}_{PY}^{G},$$

$$\mathbf{u}_{WCP} =$$

$$= \mathbf{unit}\left[max\left(PROJ - \cos 65^{0},\ 0\right)\mathbf{R}_{LOSP} + max\left(\cos 75^{0} - PROJ,\ 0\right)\mathbf{F}_{RP}\right].$$

6. ***Re-designation of landing site***
   The position vector of the landing site in the platform coordinate system, $\mathbf{R}_{LSP}$ at current time, t is computed based on the line of site vector, $\mathbf{R}_{LOSP}$:

$$\mathbf{R}_{LSP} = \mathbf{R}_{LSP}\mathbf{unit}\left(\mathbf{R}_{P} + \mathbf{R}_{LOSP}\frac{R_{XLSP} - R_{XP}}{R_{XLSP}}\right).$$

7. ***Landing point re-designator angle***
   This angle represents the angle between vectors $\mathbf{R}_{LOSP}$ and $\mathbf{u}_{XB}$ (see Fig. 8.1):

$$\mathbf{R}_{LOSP}\mathbf{u}_{XB} = |\mathbf{R}_{LOSP}||\mathbf{u}_{XB}|\cos(\mathbf{R}_{LOSP}, \mathbf{u}_{XB}) = -\sin(\theta_{LPD}),$$

$$\theta_{LPD} = -arcsin(\mathbf{R}_{LOSP}, \mathbf{u}_{XB}).$$

8. ***Position and velocity in guidance frame***
   Using the vectors defined above, one can transform the position and velocity vectors determined in the platform system into the guidance frame:

$$\mathbf{R}_{G}(t) = \mathbf{T}_{P}^{G}(\mathbf{R}_{P}(t) - \mathbf{R}_{LSP}(t)),$$
$$\mathbf{V}_{G}(t) = \mathbf{T}_{P}^{G}(\mathbf{V}_{P}(t) - \omega_{MP} \times \mathbf{R}_{P}(t)).$$

9. ***Range to the landing site***
The range from LM to the landing site is computed as

$$|\mathbf{R}_G| = \sqrt{R_{XG}^2 + R_{YG}^2 + R_{ZG}^2}.$$

10. ***Transformation matrix from platform to guidance frame***
The transformation matrix from the platform system to the guidance frame is determined as

$$\mathbf{T}_P^G(t) = \begin{bmatrix} TPG_X \\ TPG_Y \\ TPG_Z \end{bmatrix},$$

where accepting $K = 1$ for the braking phase, and $K = 0$ for the approach phase, one can erect the guidance frame for a succeeding guidance cycle pass:

$$\begin{aligned} TPG_X(t) &= \mathbf{unit}[\mathbf{R}_{LSP}(t)], \\ TPG_Y(t) &= \mathbf{unit}[\mathbf{R}_{LSP}(t) \times \mathbf{D}(t, TTT)], \\ TPG_Z(t) &= \mathbf{unit}[TPG_X(t) \times TPG_Y(t)], \end{aligned}$$

where

$$\mathbf{D}(t, TTT) = \mathbf{R}_P(t) - \mathbf{R}_{LSP}(t) -$$
$$- K(\mathbf{V}_P(t) - \omega_{MP}(t) \times \mathbf{R}_P(t))\frac{TTT}{4}.$$

Note that in Apollo P63-64 guidance algorithm the expression $(\mathbf{V}_P - \omega_M \mathbf{R}_{LSP})$ has coefficient K such that $K = 1$ if the lander on braking phase and $K = 0$ if the lander is on approach phase. This would be interpreted as the rotation of the Moon is not taken into account when the lander is close to the landing site. The vector TPG_Y is derived above in a separate subsection.

11. ***Target referenced time update***
The target referenced time can be updated as

$$TTT = TTT + t - t_{OLD},$$

and the clock-time can be assigned to t_{OLD}:

$$t_{OLD} = t.$$

The procedure of computation of the initial value of target referenced time, TTT is given below in a separate subsection.

12. ***Computation of target referenced time***
The convergence criteria is defined as

$$\Delta TTT_{\varepsilon} = \frac{|TTT|}{128}.$$

The target referenced time can be computed such that the terminal jerk constraint is satisfied, that is

$$TTT = TTT + \Delta TTT,$$

where

$$\Delta TTT = -\frac{d_1 + d_2}{d_3},$$

and

$$d_1 = J_{ZTG} TTT^3 + 6A_{ZTG} TTT^2,$$

$$d_2 = (18V_{ZTG} + 6V_{ZG})TTT + 24(R_{ZTG} - R_{ZG}),$$

$$d_3 = 3J_{ZTG} TTT^2 + 12A_{ZTG} TTT + 18V_{ZTG} + 6V_{ZG}.$$

If $|\Delta TTT| > \Delta TTT_{\varepsilon}$, then one can repeat the process with new TTT, otherwise this new TTT is used further to compute the commanded thrust acceleration. Here ΔTTT_{ε} is a user-defined quantity.
13. ***Commanded thrust acceleration***
First, let's compute the ratio of the predicted target referenced time to the current target referenced time:

$$T_P = \frac{TTT + t_{LEAD}}{TTT},$$

where t_{LEAD} is transport delay. Then the acceleration to be commanded at the current time TTT and realized at the predicted time T_P can be computed as follows

$$\mathbf{A}_{CG} = 12\frac{T_1}{TTT^2}(\mathbf{R}_{TG} - \mathbf{R}_G) + 6\frac{T_2}{TTT}\mathbf{V}_{TG} + 6\frac{T_3}{TTT}\mathbf{V}_G + T_4\mathbf{A}_{TG},$$

where

$$T_1 = 3T_P^2 - 3T_P,$$
$$T_2 = 4T_P^2 - 3T_P,$$
$$T_3 = 2T_P^2 - T_P,$$
$$T_4 = 6T_P^2 - 6T_P + 1.$$

14. ***Commanded unit thrust vector***
The commanded unit thrust vector is defined directly from the commanded thrust acceleration vector as follows:

$$\mathbf{A}_{CP} = (\mathbf{T}_P^G)^T \mathbf{A}_{FCG},$$

$$\mathbf{A}_{FCP} = \mathbf{A}_{CP} - \mathbf{G}_{MP},$$

$$\mathbf{U}_{FCP} = \mathbf{unit}[\mathbf{A}_{FCP}].$$

15. ***Commanded unit window vector***
The computation of the commanded window vector $\mathbf{U}_{WCP}$ is based on the commanded unit thrust vector, the line of site and the elements of the matrix $\mathbf{T}_P^G$ (see Fig. 3.3 of Chap. 3):

$$\mathbf{u}_{XB} = \mathbf{U}_{FCP},$$

$$\mathbf{F}_{RP} = \mathbf{unit}\left(\mathbf{u}_{XB} \times \mathbf{T}_{PY}^G\right),$$

$$PROJ = (\mathbf{R}_{LOSP} \times \mathbf{u}_{XB})\, \mathbf{T}_{PY}^G,$$

$$\mathbf{U}_{WCP} = \mathbf{unit}\left[\mathbf{D}_1 + \mathbf{D}_2\right],$$

where

$$\mathbf{D}_1 = max\left(PROJ - \cos 65^0,\ 0\right) \mathbf{R}_{LOSP},$$

$$\mathbf{D}_2 = max\left(\cos 75^0 - PROJ,\ 0\right) \mathbf{F}_{RP}.$$

Reference

1. Klumpp AR (1971) "Apollo Lunar-Descent Guidance", Apollo Guidance Navigation Control, R-695. Massachusetts Institute of Technology, MIT Instrumentation Laboratory Report, Cambridge, MA

Chapter 9
P66 Lunar-Landing Guidance

As mentioned in Chap. 2, the Program 66 (P66) consists of the vertical and horizontal channel design for guiding the LM on the last phase of the trajectory. The formulas of this P66 guidance program and the commanded body frame matrix formulas are described below.

9.1 P66 Horizontal Guidance Formula

The P66 horizontal guidance algorithm is processed once every 2 s, nulls the horizontal velocity component relative to the lunar surface and is given by the following formulas [1]:

$$
\begin{aligned}
&AFCP_X = GM \\
&AFCP_Y = LIMIT[-\frac{1}{5}(V_{PY} - VMOONP_Y) - 0.4AFCP_Y,\ (20/57.3)GM] \\
&AFCP_Z = LIMIT[-\frac{1}{5}(V_{PZ} - VMOONP_Z) - 0.4AFCP_Z,\ (20/57.3)GM],
\end{aligned}
$$

where **AFCP** is the thrust acceleration command, GM is the lunar gravitational acceleration, $\mathbf{V_P}$ is the current velocity. **VMOONP** is the lunar-surface velocity in the platform frame.

© The Author(s), under exclusive license to Springer Nature Switzerland AG 2026

D. M. Azimov and R. H. Bishop, *Apollo-Class Targeting and Guidance for Lunar Powered Descent and Precision Landing*, Springer Aerospace Technology,
https://doi.org/10.1007/978-3-031-91088-3_9

9.2 P66 Vertical Guidance Formula

The P66 vertical (ROD) guidance algorithm is processed every second, controls altitude rate to the reference value by throttling the DPS. The scalar equivalent of the ROD equation is given as [1]

$$AFCP = \frac{-\frac{1}{1.5}(V_{PX} - V_{CPX)-G_{PX}}}{CBP_{XX}},$$

where $AFCP$ is the thrust acceleration command for the Throttle routine, V_{PX} is the extrapolated velocity, V_{CPX} is the commanded vertical velocity updated each pass from ROD increment commands. CBP_{XX} is the direction cosine of the XB-axis and XP-axis to correct for the non-vertical orientation of the thrust vector. Note that the body-frame unit vectors expressed in platform coordinates are the row vectors $CBP_X,\ CBP_Y$ and CBP_Z of the matrix CBP [1].

9.3 Commanded Body Frame Matrix in Attitude Routine

The guidance and navigation inputs to this routine (ATT) consist of the unit thrust command, unit window command and the thrust-acceleration measurement. The ATT processes the thrust-acceleration measurement in a thrust direction filter to estimate a unit thrust vector with respect to the LM body axes. Using the estimated unit thrust vector and the input commands, the ATT produces a commanded-body frame matrix, $CCBP$. This matrix is the transformation matrix from the platform to the commanded-body coordinates. Therefore, it can be expressed in terms of the IMU gimbal angles which would place the body axes in the commanded directions and the commanded gimbal angles can be extracted from it. The $CCBP = CCBP(i, j),\quad i, j = 1, 2, 3$ is a product of three matrices that correspond to the gimbal angles $\phi_1,\ \phi_2$ and ϕ_3 of rotations about the three gimbal axes $X,\ Y$ and Z [1]:

$$\begin{aligned}
CCBP(1, 1) &= \cos\phi_3 \cos\phi_2\\
CCBP(1, 2) &= \sin\phi_3\\
CCBP(1, 3) &= -\cos\phi_3 \sin\phi_2\\
CCBP(2, 1) &= -\cos\phi_1 \sin\phi_3 \cos\phi_2 + \sin\phi_1 \sin\phi_2\\
CCBP(2, 2) &= \cos\phi_1 \cos\phi_3\\
CCBP(2, 3) &= \cos\phi_1 \sin\phi_3 \sin\phi_2 + \sin\phi_1 \cos\phi_2\\
CCBP(3, 1) &= \sin\phi_1 \sin\phi_3 \cos\phi_2 + \cos\phi_1 \sin\phi_2\\
CCBP(3, 2) &= -\sin\phi_1 \cos\phi_3\\
CCBP(3, 3) &= -\sin\phi_1 \sin\phi_3 \sin\phi_2 + \cos\phi_1 \cos v2,
\end{aligned}$$

The commanded gimbal angles are given as

$$CGA_X = TAN^{-1}\left[-\frac{CCBP_{ZY}}{CCBP_{YY}}\right],$$

$$CGA_Y = TAN^{-1}\left[-\frac{CCBP_{XZ}}{CCBP_{XX}}\right],$$

$$CGA_Z = TAN^{-1}\left[-\frac{CCBP_{XY}}{\sqrt{1 - CCBP_{XY}^2}}\right].$$

9.4 Concluding Remarks

In this and previous chapters the Apollo ground-based targeting and two on-board algorithms for the powered lunar descent and landing maneuver as well as the attitude routine have been described. The targeting program includes necessary derivations, approach-phase and braking-phase targeting algorithms together with constraints, inputs and outputs. Understanding and proper interpretation of these algorithms are important in constructing and executing the entire flight program for a soft landing on the Moon, and on any celestial body in general. The remaining algorithms are subject to further studies. Besides that, some necessary derivations and explanations, as well as inputs and outputs of the algorithms have been presented. The iterative procedures which take place in the braking-phase targeting algorithm, and certain aspects of both targeting and guidance procedures have been presented in the most possible explicit forms. Some minor aspects of the Apollo algorithms, such as the specific values of some coefficients are still remaining unknown and subject to trials.

Next part of this monograph presents the modifications of the Apollo targeting and guidance programs. The corresponding studies include the explicit formulation of the braking-phase targeting, interpretation and explanation of the on-board algorithms. In particular,

- development of the simulation model utilizing the original Apollo targeting and guidance parameters,
- utility of the quartic properties,
- development of explicit on-board targeting by utilizing analytical thrust arc solutions and eliminating the uncertainties and iterations, which took place in the Apollo targeting and guidance codes,
- simulations on explicit on-board targeting according to the closed-form analytical solutions,
- development of the concepts re-targeting procedures, and
- description of the ways of improving the landing accuracy based on the coordinate systems and transformations used in Apollo guidance are presented.

Reference

1. Klumpp AR (1971) "Apollo Lunar-Descent Guidance," Apollo Guidance Navigation Control, R-695. Massachusetts Institute of Technology, MIT Instrumentation Laboratory Report, Cambridge, MA

Chapter 10
Simulations and Comparison Analyses

10.1 Approach to Simulations

In this chapter, the comparison analysis of the trajectory reconstruction and post-flight data for Apollo 11 and 12 mission are presented [1, 2]. The Apollo targeting and guidance parameters are generic and do not reflect any specific mission parameters with predetermined landing sites. Therefore, the initial entry orbit parameters, the parameters of descent propulsion system and landing site coordinates with corresponding epochs have been specified in the simulations according to a specific mission for comparison purposes. The comparisons of thrust and attitude profiles, the time histories for altitude, current range to landing site and velocity vector magnitude, and some data points for mass, thrust, flight path angle and central angle are presented. The terminal descent phase is almost vertical, and typically lasts about 30 s and not considered in these simulations. Given the initial conditions, the landing site coordinates and the nominal landing time, the targeting program calculates the state vector at target points of both braking and approach phases. The constant matrix $\mathbf{T}_P^I$ is computed at perilune of the entry orbit before the maneuver starts. The matrix $\mathbf{T}_I^L$ between the inertial and local systems is computed according to the Apollo Project Standard Coordinate Systems [3]. The target states computed by the targeting program are used in guidance to satisfy the terminal conditions of each phase. In the simulations, it is assumed that the braking phase with CT (93% of nominal F_{max}) starts at the perilune of the entry orbit. This point is the terminal point of the trim phase with minimum thrust (10% of nominal F_{max}) [1, 4]. According to the definitions of the coordinate systems, the perilune of the entry orbit, the Moon COM and the nominal landing site at the nominal landing time lie in the same plane which coincides with the plane of the entry orbit. Also, this plane coincides with or parallel to the planes of the platform and guidance coordinate systems if it is assumed that the lander lands at the nominal landing site at the nominal landing time [5]. As the platform system is non-rotating and the guidance system's origin moves together with the Moon's surface, these two planes will slightly deviate during the landing

© The Author(s), under exclusive license to Springer Nature Switzerland AG 2026

D. M. Azimov and R. H. Bishop, *Apollo-Class Targeting and Guidance for Lunar Powered Descent and Precision Landing*, Springer Aerospace Technology,
https://doi.org/10.1007/978-3-031-91088-3_10

maneuver. This means that although the trajectory plane remains close to the plane of the platform system, it will have a curvilinear character due to the motion of the landing site as a result of Moon's rotation. It is assumed that the lander is already in the entry orbit, and no plane change is considered in this work.

10.2 Comparison with Apollo 11 Post-Flight Data

A series of simulations have been conducted to compare the profiles of various important parameters to the Apollo 11 mission trajectory data [1]. The trajectory reconstruction and post-flight analysis of the Apollo 11 mission are given in Refs. [1, 6]. Our simulation results are compared primarily to the respective profiles provided in these documents. The AP and BP targeting constraints, inputs and outputs are described in Ref. [7].

The Apollo 11 landing site is considered to be Site 2 with the coordinates 0°42′50″ latitude North, 23°42′28″ longitude East [8]. It is located on the east central part of the Moon in southwestern Mar Tranquillitatis. The actual landing site was 0°41′15″ latitude North and 23°26′00″ longitude East, which was located at 6870 meters from the center of the landing ellipse. The landing time was at 3:17 PM Eastern Standard Time (EST). The inputs of AP and BP targeting programs are provided in Tables 10.1 and 10.2. The corresponding target states for these phases as the outputs of the targeting programs are presented in Tables 10.3 and 10.4. The values of altitude, altitude rate and range at high and low gates on the trajectory are given in Table 10.5 [1].

The commanded and actual thrust, the pitch angle (the angle determined by using the X component of the thrust vector), the time histories for selenocentric altitude, current range to landing site and velocity vector magnitude, and the data points for mass, thrust, flight path angle and central angle at some points of the trajectory are compared to the same parameters of Apollo 11 mission. It should be noted here that

Table 10.1 Inputs of AP targeting program of Apollo 11

Terminal target referenced time	$TAPF = -10$ s
Midpoint target referenced time	$TAPM = -76$ s
Initial target referenced time	$TAPI = -186$ s
The landing site must be approached with this slope	$\gamma = 15.943$ deg
Effective P66 horizontal time constant	$\tau = 8$ s
Terminal altitude	$R_{XAPFG} = 30$ m
Terminal altitude rate	$V_{XAPFG} = -1$ m/s
The X-components of the position and velocity vectors at the midpoint	$R_{XAPMG} = 150.0$ m $V_{XAPMG} = -5.0$ m/s
The X-component of the initial position vector	$R_{XAPIG} = 2231.0$ m

Table 10.2 Inputs of BP targeting program for of Apollo 11

Magnitude of the gravitational acceleration on the Moon surface	$G_0 = 1.62$ m/s^2
Mean radius of the Moon	$R_M = 1736000$ m
Altitude of the landing site (wrt surface)	0.0
Selenocentric latitude of the landing site	0.706944 deg
Longitude of the landing site	23.707777 deg
Altitude of the perigee of the entry orbit	$h_p = 15240$ m
Altitude of the apogee of the entry orbit	$h_a = 111120$ m
Central angle between the landing site vector and lander's position vector at t_0 measured in the trajectory plane (platform system)	$\theta_0 = 14.85$ deg
P64 initial state	$R_{XAPIG} = 2231$ m
Terminal pitch angle constraint	$\theta_{PBRF} = 59.93$ deg
Duration of trim phase starting from t_{IG} to t_I	$\Delta t_{TRIM} = 26$ s
Duration of ballistic phase starting from t_0 to t_{IG}	$\Delta t_{BP} = 630$ s
Duration of ullage	$\Delta t_{ULLAGE} = 7.5$ s
Time interval between throttle down point (TDP) and nominal landing time t_{LAND}	$\Delta t_{LTDI} = 30$ s
Required time duration between TDP and BR terminal point	$\Delta t_{THR} = 120$ s
Nominal landing time measured from t_I	$t_{LAND} = 721$ s
Initial target referenced time at first guidance call at t_I	$T = -664$ s
Target referenced time at BR terminal point	$TBRF = -60.0$ s
Initial mass of the lander	$M_0 = 15500$ kg
Magnitude of exhaust velocity of gas particles in DPS	$V_E = 2955.889$ m/s
DPS nominal maximum thrust	$F_{max} = 46706$ N
Convergence criterion for relative terminal state errors	$\epsilon_{SQ} = 5 \times 10^{-2}$

Table 10.3 State at BP target point of Apollo 11

$RBRTG$:	-182.488563846195	0.00	-3756.37475891394
$VBRTG$:	-43.6013002873879	0.00	-13.9016757316038
$ABRTG$:	-0.137344058309065	0.00	-2.77861802715640
$JBRTG$:	$-2.16055626909E-3$	0.00	$-3.5287740590E-3$
$SBRTG$:	$-6.13892113488E-5$	0.00	$-1.4977323011E-5$

Table 10.4 State at AP target point of Apollo 11

$RAPTG$:	16.8377946103335	$-6.93942699916E-5$	-0.372361050101432
$VAPTG$:	-1.69062338285702	$-2.12367035711E-5$	-0.331483078707317
$AAPTG$:	$-8.657346653E-2$	$-4.41470975307E-6$	-0.160328711692761
$JAPTG$:	$-3.524800444E-3$	$-4.91681687032E-7$	$-1.38717264350E-3$
$SAPTG$:	$-6.772438626E-6$	$-1.00421423450E-8$	$-1.31955103240E-4$

Table 10.5 Data points: Apollo 11 data and the simulation results

Trajectory point, parameter	Apollo 11 nominal (ft, ft/s)	Authors' results (ft, ft/s)
High gate		
Altitude	7600.0	7468.60
Range	26000.0	26676.00
Altitude	512.1	512.00
Altitude rate	−16.01	−16.78
Low gate		
Altitude	500.0	500.22
Range	2000.0	1886.20
Start of vertical descent		
Altitude	150.0	149.93
Altitude rate	3.0	2.26

the actual and precise target states and other targeting parameters used in the Apollo 11 and other Apollo missions are unknown to the authors.

The comparison of profiles of commanded and actual thrust data on AP and BP of Apollo 11 given in Fig. 10.1 shows that although the initial values of actual thrust are different, the time histories of thrust profiles are very close or almost identical [1]. The thrust magnitude is switched from 93% to about 57% by the guidance program at almost the same target referenced times. This comparison indicates that the target states computed by the targeting program are close to those used in the Apollo pre-mission targeting computations. Qualitative comparison in Fig. 10.2 shows that the profiles of the pitch angle from vertical are also very close [1]. Note that the change in the angle is almost the same at most points on the trajectory.

The post-flight analysis of mission data made it possible to construct various qualitative relationships between main parameters of the trajectory. One such relationship is the commanded thrust percentage versus horizontal velocity (shown in Fig. 10.3). It can be seen that the behavior of the commanded thrust with respect to the horizontal velocity obtained by the authors is almost the same as in the Apollo 11 mission. The results are consistent with commanded and actual thrust data reflected in Fig. 10.3. Another comparison of parameters resulted from the simulations and Apollo 11 mission data can be analyzed with respect to the relationship between the altitude and the current range to the landing site. Due to the difference in actual locations of the origin of the platform and guidance coordinate systems, it was more convenient to compute the altitude in the platform system during the early stages of the trajectory. But this system does not provide a very accurate altitude when the lander is close to the landing site because the trajectory plane slightly deviates from the actual immediate plane of motion. In this case, it is more convenient and accurate to measure the altitude in the guidance system. The relationships between the altitude and the range to the landing site on the approach phase, presented in

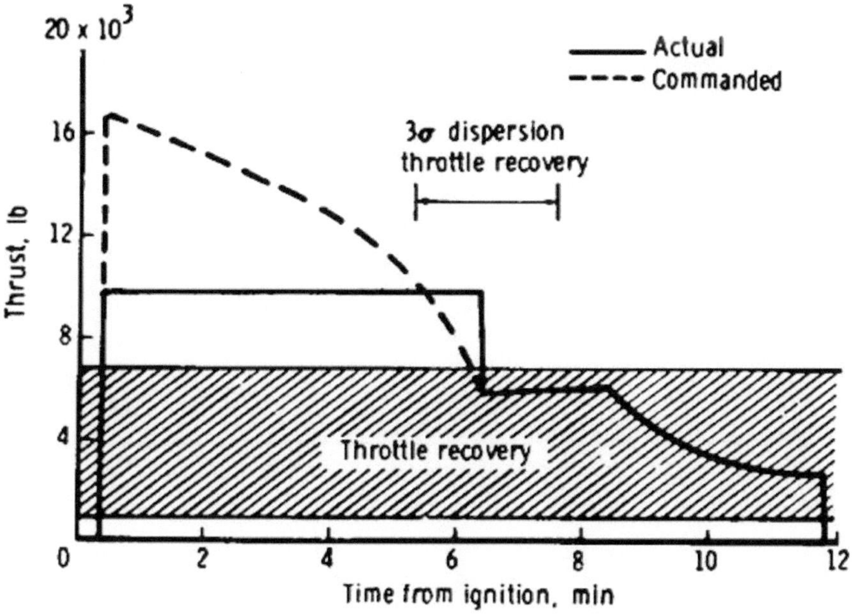

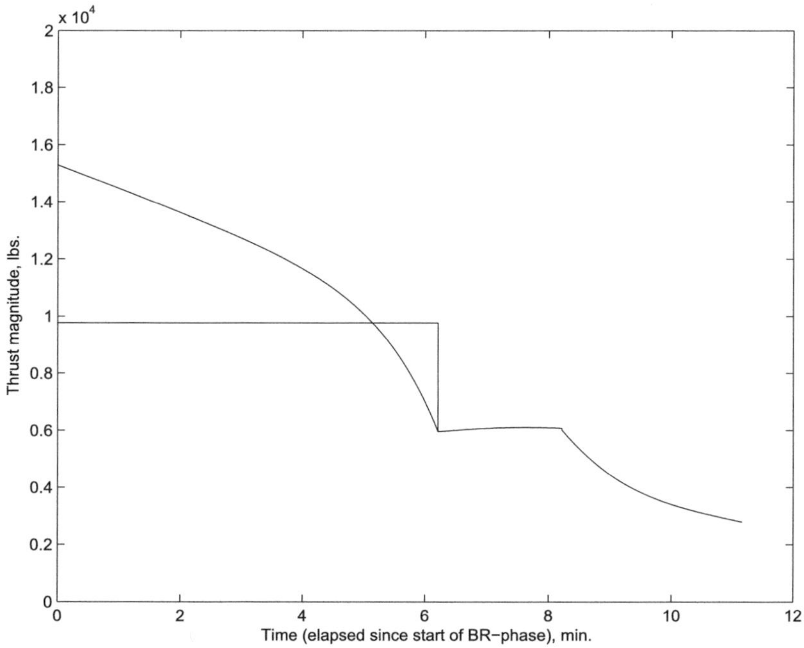

Fig. 10.1 BP and AP: actual and commanded thrust profiles of Apollo 11 Ref. [1] (above) and from the simulations (below)

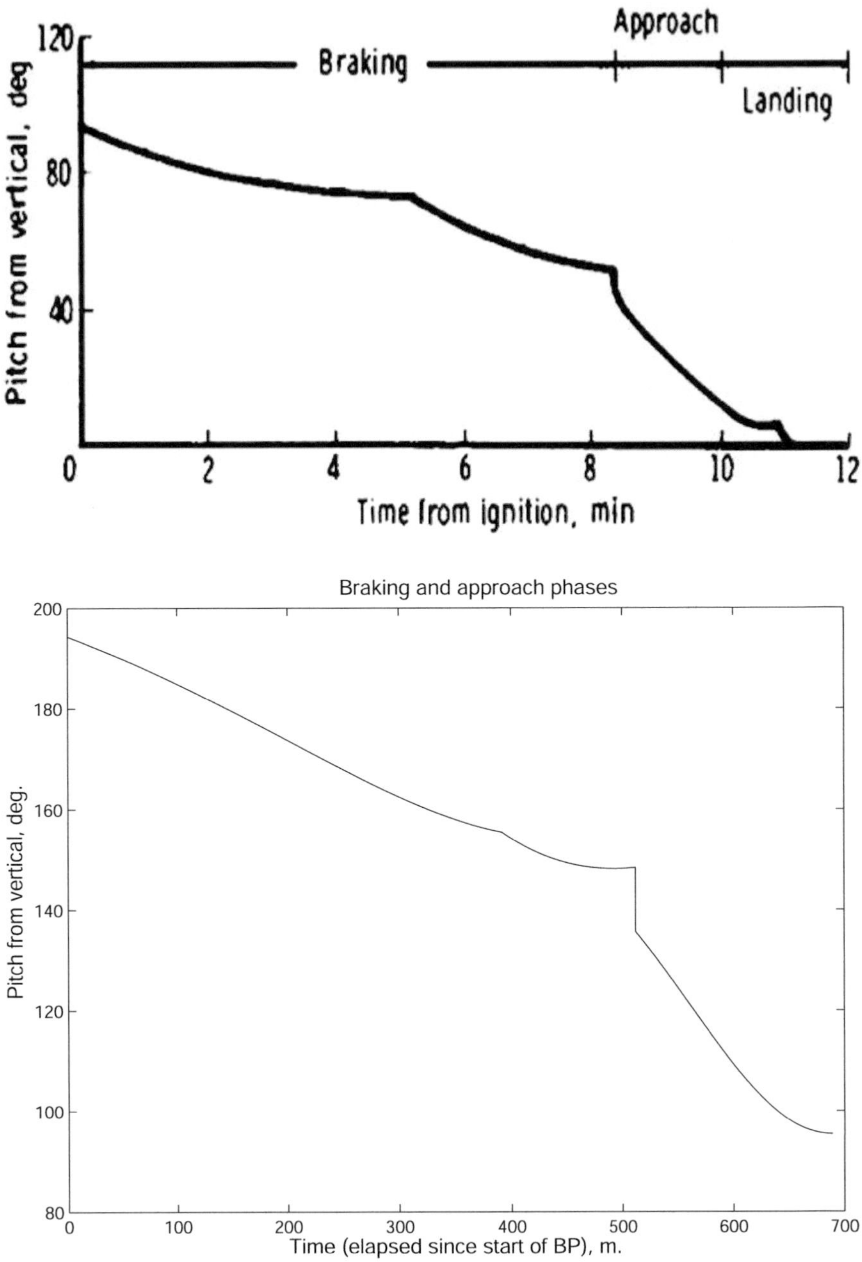

Fig. 10.2 BP and AP: LM's pitch profile of Apollo 11 [1] (above) and from the simulations (below)

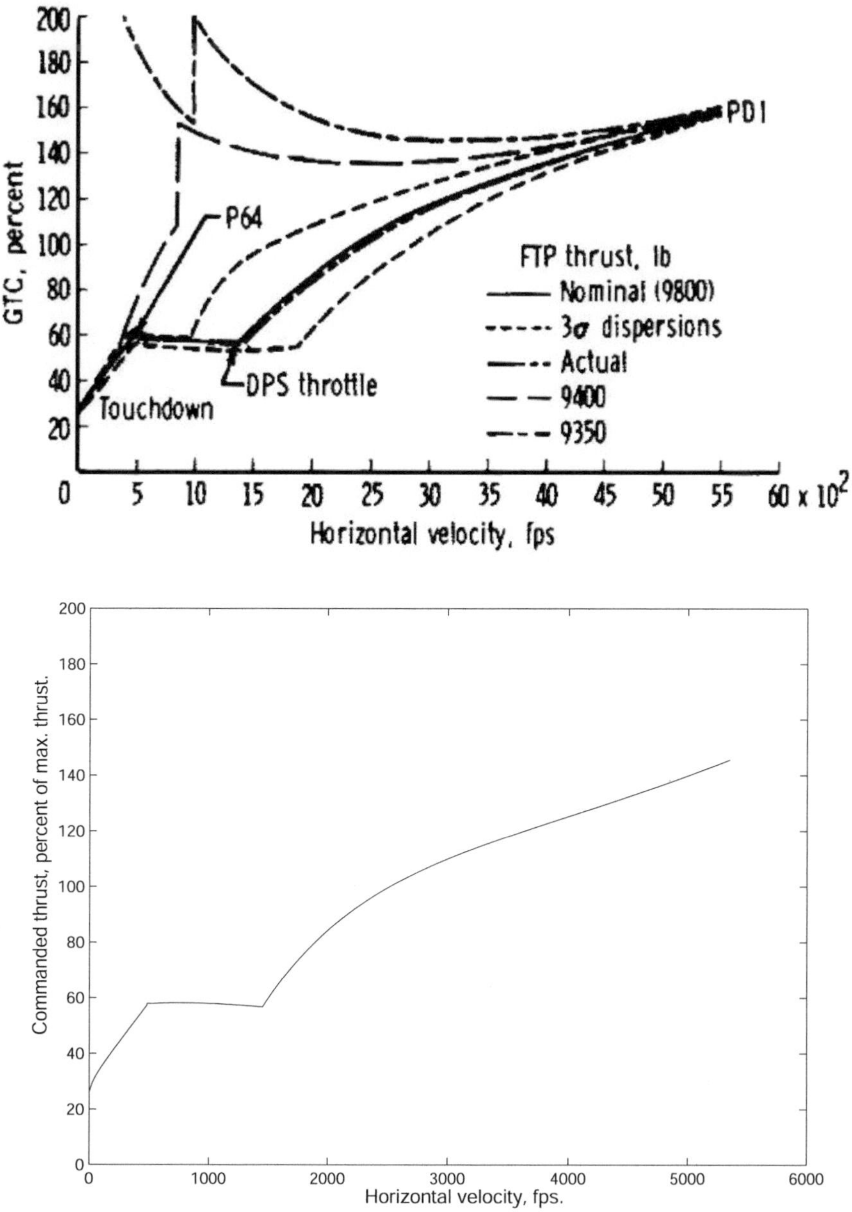

Fig. 10.3 Commanded thrust versus horizontal velocity profile of Apollo 11 [1] (above) and from the simulations (below)

Fig. 10.4, are constructed by measuring these distances in the guidance system. The plots show almost identical behaviors in both cases. Once more, this closeness of results indicates that the computed target states of the approach phase and other trajectory parameters are consistent with the Apollo 11 post-flight trajectory data. It is also important to consider the trajectory plot in the guidance frame and the change in mass and current selenocentric altitude since the beginning of the maneuver. These plots are presented in Fig. 10.5. Although the main literature does not contain the time histories of mass or trajectory data in the guidance reference plane, it is possible to analyze the corresponding plots in the context of the trajectory parameters considered above. The trajectory plot shows how the lander is approaching and how close it is to the landing site at any given time. The mass profile shows the fact that the thrust and the mass-flow rate are constant on BP-CT-arc, and has a noticeable change in behavior at the throttle recovery point (the point where the thrust magnitude is switched from 93% to 57%). The high gate is close to the starting point of the approach phase. The low gate point is located close to a middle point of the approach phase. Table 10.5 demonstrates the closeness of the trajectory data of the simulations and the Apollo 11 post-flight data at selected points and its consistency with other trajectory data analyzed above [1].

10.3 Comparison with Apollo 12 Post-Flight Data

Simulations and comparison analysis of profiles of various important parameters have also been conducted with the Apollo 12 post-mission trajectory data [4, 9]. The studies on numerical trajectory optimization on the braking phase of Apollo 12 and their comparison with real mission data are provided in [9]. The simulation results are compared to those given in Refs. [4, 9]. The inputs of AP and BP targeting programs of Apollo 12 are provided in Tables 10.6 and 10.7. The target states for these phases are provided in Tables 10.8 and 10.9. The trajectory date points and the comparison of the simulation results with those of Apollo 12 are given in Table 10.10.

The thrust profile on BP is presented in Fig. 10.6. The DPS thrust level used in [9] is little higher than that used in our simulations. In Ref. [9], the maximum thrust level is 93.8% and in the simulations, the authors used 93% which corresponds to the real time data presented in Ref. [4]. This resulted in the difference of initial values of actual thrust. However, the time histories of thrust profiles are very close. The actual thrust magnitude is switched from 93% to about 57% by the guidance program at almost the same time compared to the Apollo 11 mission data. As one would expect, correspondingly, that the mass profile obtained by the authors for the BP and AP is similar to those of Apollo 11 (see Fig. 10.7). The point at which the line changes is where the throttle recovery occurs. During the last 120 s of the BP and for the rest of the trajectory, the actual and commanded thrusts coincide [5]. The comparison indicates that the target states computed by the targeting program for the braking phase are close to those used in the pre-mission targeting computations. Comparison of the pitch profile of Apollo 12 given in Fig. 10.8 with the authors'

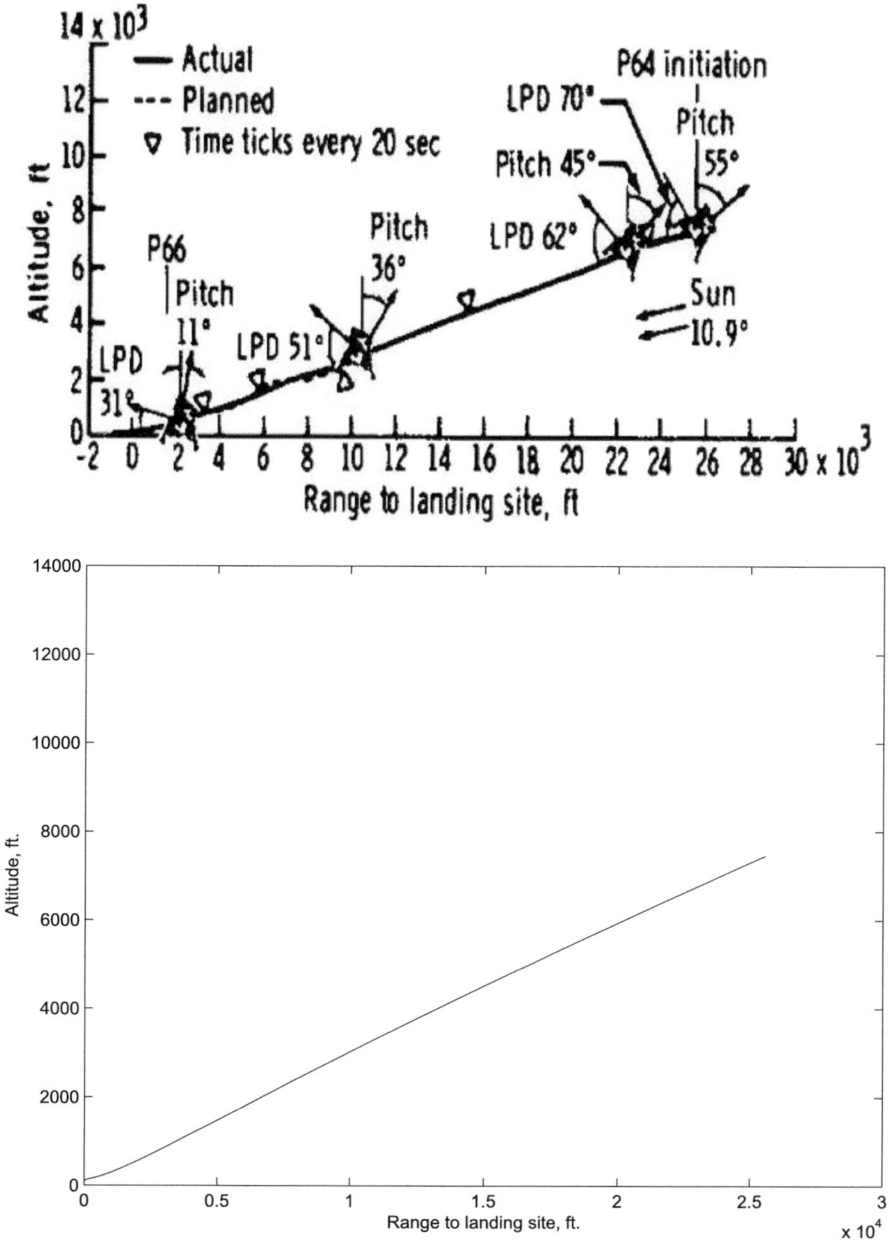

Fig. 10.4 Approach phase: Altitude versus range to landing site profile of Apollo 11 [1] (above) and from the simulations (below)

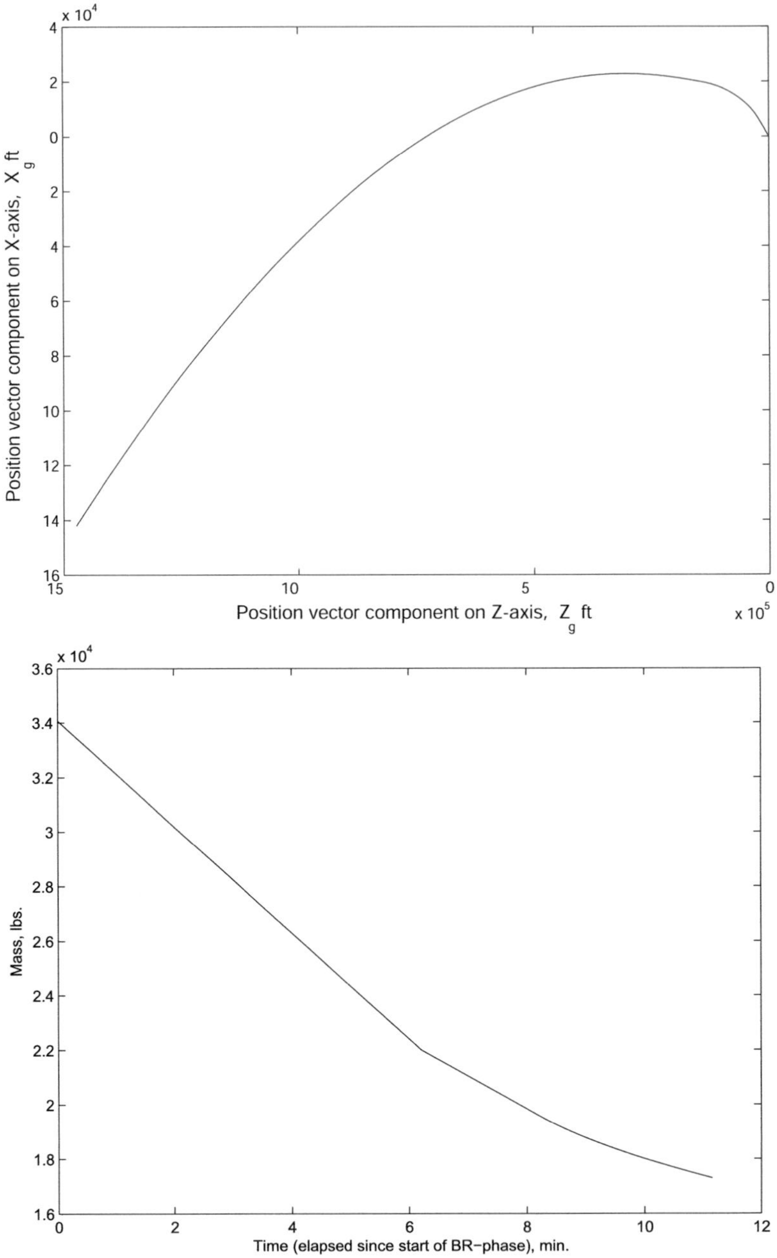

Fig. 10.5 Braking and approach phases: LM trajectory in guidance frame (above) and mass versus time (below)

Table 10.6 Inputs of AP targeting program of Apollo 12

Terminal target referenced time	$TAPF = -11$ s
Midpoint target referenced time	$TAPM = -77$ s
Initial target referenced time	$TAPI = -188$ s
The landing site must be approached with this slope	$\gamma = 16.1$ deg
Effective P66 horizontal time constant	$\tau = 8$ s
Terminal altitude	$R_{XAPFG} = 29.9$ m
Terminal altitude rate	$V_{XAPFG} = -1$ m/s
The X-components of the position and velocity vectors at the midpoint	$R_{XAPMG} = 150.0$ m $V_{XAPMG} = -5.0$ m/s
The X-component of the initial position vector	$R_{XAPIG} = 2231.0$ m

Table 10.7 Inputs of BP targeting program of Apollo 12

Magnitude of the gravitational acceleration on the Moon surface	$G_0 = 1.62\ m/s^2$
Mean radius of the Moon	$R_M = 1736000$ m
Altitude of the landing site (wrt surface)	-1.3164 m
Selenocentric latitude of the landing site	-3.036 deg
Longitude of the landing site	-23.4180 deg
Altitude of the perigee of the entry orbit	$h_p = 14741.920$ m
Altitude of the apogee of the entry orbit	$h_a = 115379.60$ m
Central angle between the landing site vector and lander's position vector at t_0 measured in the trajectory plane (platform system)	$\theta_0 = 15.43$ deg
P64 initial state	$R_{XAPIG} = 2231$ m
Terminal pitch angle constraint	$\theta_{PBRF} = 59.65$ deg
Duration of trim phase starting from t_{IG} to t_I	$\Delta t_{TRIM} = 26$ s
Duration of ballistic phase starting from t_0 to t_{IG}	$\Delta t_{BP} = 630$ s
Duration of ullage	$\Delta t_{ULLAGE} = 7.5$ s
Time interval between throttle down point (TDP) and nominal landing time t_{LAND}	$\Delta t_{LTDI} = 30$ s
Required time duration between TDP and BR terminal point	$\Delta t_{THR} = 120$ s
Estimated clock-time of landing measured from t_I	$\Delta t_{LAND} = 700$ s
Initial target referenced time at first guidance call at t_I	$T = -664$ s
Target referenced time at BR terminal point	$TBRF = -50.0$ s
Initial mass of the lander	$M_0 = 15121$ kg
Magnitude of exhaust velocity of gas particles in DPS	$V_E = 2955.889$ m/s
DPS nominal maximum thrust	$F_{max} = 43903.734$ N
Convergence criterion for relative terminal state errors	$\epsilon_{SQ} = 5 \times 10^{-2}$

Table 10.8 State at BP target point of Apollo 12

$RBRTG$:	231.853774335062	0.00	−3734.35694074137
$VBRTG$:	−43.6910905015877	0.00	15.0163134638275
$ABRTG$:	−0.164980295097671	0.00	−2.64136588507095
$JBRTG$:	$-1.810895358559421E-3$	0.00	$-3.162377950711337E-3$
$SBRTG$:	$-6.493408197035151E-5$	0.00	$-1.420401605353727E-5$

Table 10.9 State at AP target point of Apollo 12

$RAPTG$:	14.9464375909053	$-1.55574878550E-4$	−0.707381629871073
$VAPTG$:	−1.79097209208152	$-4.33331774382E-5$	−0.464757095453972
$AAPTG$:	$-9.16766147237E-2$	$-8.20736498961E-6$	−0.154928462327103
$JAPTG$:	$-3.62903517654E-3$	$-8.35743759381E-7$	$-1.16275938982E-3$
$SAPTG$:	$-9.40326791231E-6$	$-1.62944853479E-8$	$-1.22075682623E-4$

Table 10.10 Data points: Apollo 12 data and the simulation results. Note that "Throttle down" is TCR

Event	Ignition	Full throttle	Throttle down	High gate
Time since ignition	0 0	26.0 26.0	388.0 391.9	508.0 511.9
Inertial pos. (m)	1751100	1750530	1743230	1737840
	1750742	1750742	1742254.63	1738281
Inertial vel. (m/s)	1695.21	1687.21	426.61	132.452
	1696.66	1689.11	429.80	148.47
Flight path angle (d)	0.0 0.0	−0.026 0.0047	−3.987 −2.783	−21.341 −15.19
Central angle (d)	0.0 0.0	1.439 0.0	14.04 15.429	15.514 15.18
Vehicle weight (kg)	15121.4	15079.5	9668.90	8636.40
	15121.4	15082.38	9668.90	8636.40

simulations shows that the pitch angle profiles are very close qualitatively [9]. It should be noted that the Apollo's value of the final pitch angle on the BP is little higher compared to the simulation results. Note that the initial altitude of the starting point of the braking phase with maximum thrust (perilune of the entry orbit) used in the simulations corresponds to Ref. [4], and it is lower than that used in Ref. [9]: $h_p = 14741.92$ m to $h_p = 15240$ m. The profiles of the magnitude of the velocity vector computed in the platform frame for the BP given in Fig. 10.9 and those of the authors are almost the same. The lines in the plots change at almost the same time when the thrust changes from 93% to about 57% (the TCR point).

Interesting comparison of the relationships between the current altitude and the current range to the current landing site is presented in Fig. 10.10. The initial altitudes in these figures differ slightly due to the different initial altitudes mentioned above.

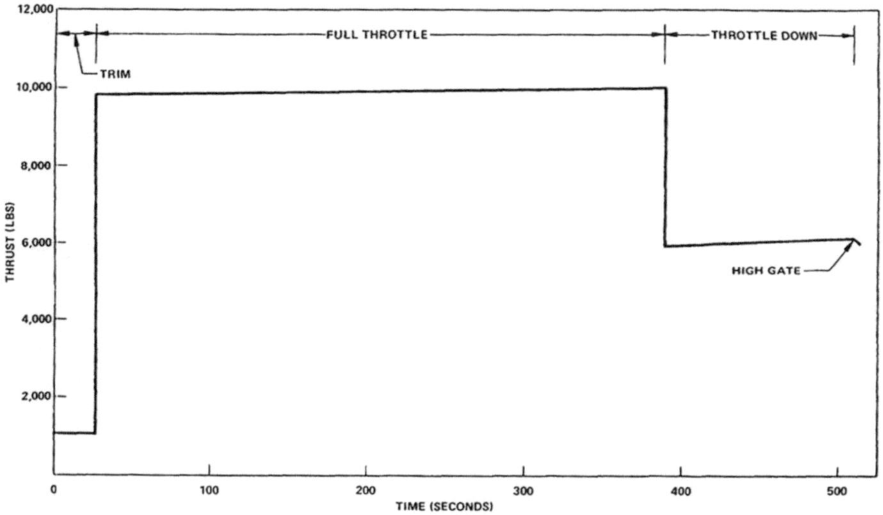

Fig. 10.6 Reference [9]. Braking phase: thrust profile

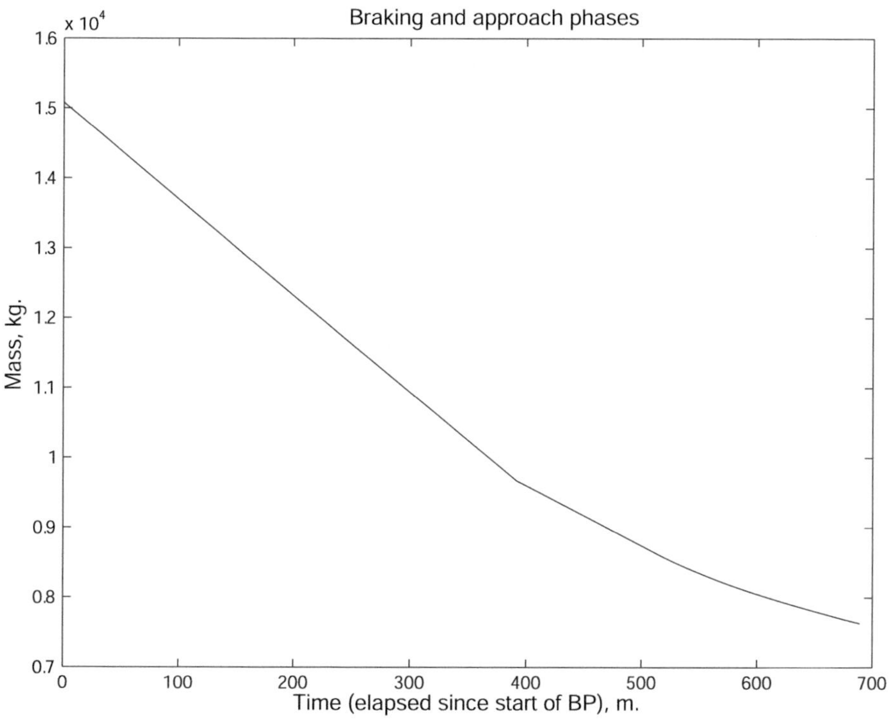

Fig. 10.7 Mass versus time: BP and AP (from the authors' simulations)

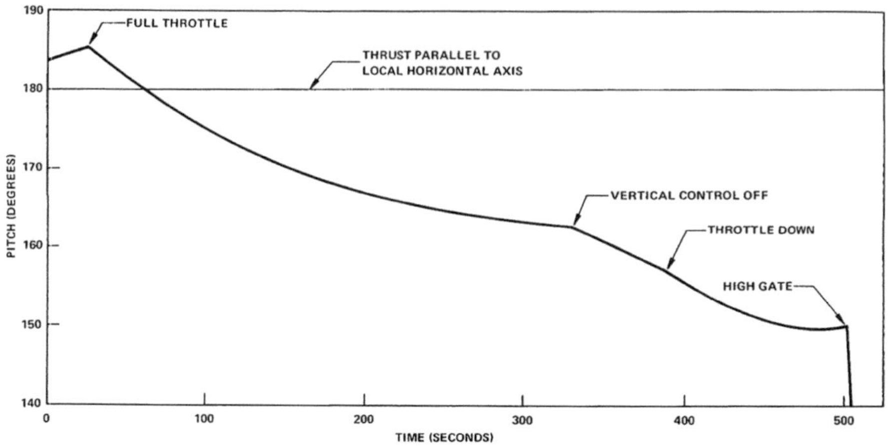

Fig. 10.8 Reference [9]. Braking phase: pitch profile

The plots show the same qualitative and quantitative behavior for the altitude and the range. As in the case of comparison with Apollo 11 mission data, it is also important to consider the trajectory in the guidance frame and also the change of mass versus time. The trajectory plot shows that at final time of the AP the lander is almost above the landing site.

The best estimate of the landing site coordinates for the Apollo 12 mission was $-3.036°$ latitude South, $-23.418°$ longitude East, and at -1.3164 m altitude [4]. The values of time, inertial position and velocity vectors, flight path angle, central angle and mass are compared for various points of the Apollo and simulated trajectories [9]. These points are the ignition, initial point of the BP which starts with maximum thrust, throttle control recovery point and the initial point of the approach phase (high gate) (see Table 10.10). According to Ref. [9], a typical entry orbit's perilune is 15186.4 m and its apolune is 111120 m. However, the entry orbit used by the authors is the same as the nominal orbit considered for the Apollo 12 mission, that is the perilune is 14741.92 m and the apolune is 115379.60 m [4]. Lower altitude of the starting point may have caused some differences shown in the Table 10.10. Another reason for these differences, taking into account the closeness of the comparison parameters, is that in the simulations, the braking phase with maximum thrust starts at the perilune. But according to Ref. [9], this phase starts at a point of the entry orbit located before the perilune with a higher altitude than the one used in the simulations. Also, it is assumed that the position vector is not changed during the trim phase [5]. The altitude versus time profile of Apollo 12 for the range from the low gate to landing is presented in Fig. 10.11 [4]. The authors' simulations of the similar profile for the BP and AP presented in Fig. 10.12 show very close results compared to those shown in Fig. 10.11.

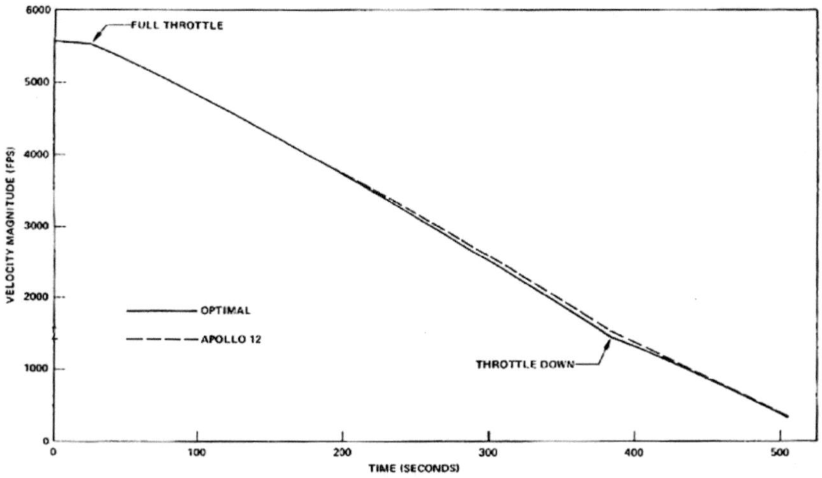

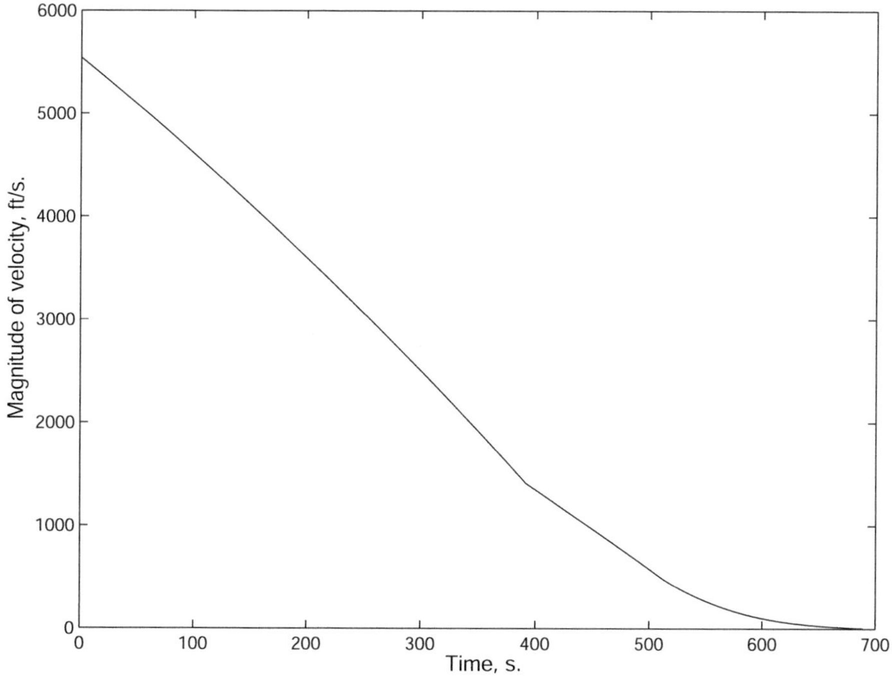

Fig. 10.9 Velocity magnitude versus time: BP [9] (above), and BP and AP (from the simulations (below)

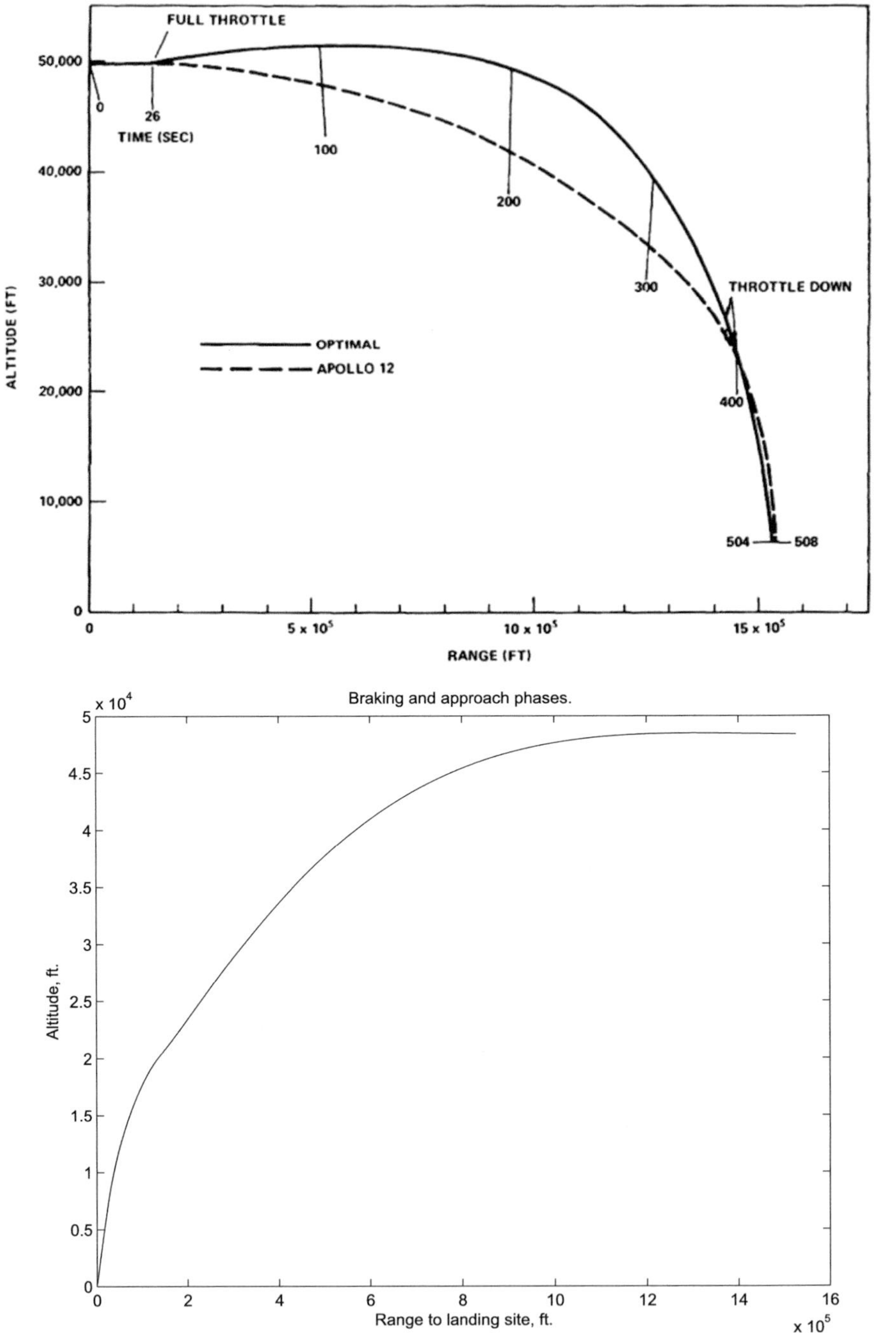

Fig. 10.10 Altitude versus range: BP [9] (above), and BP and AP (from the simulations (below)

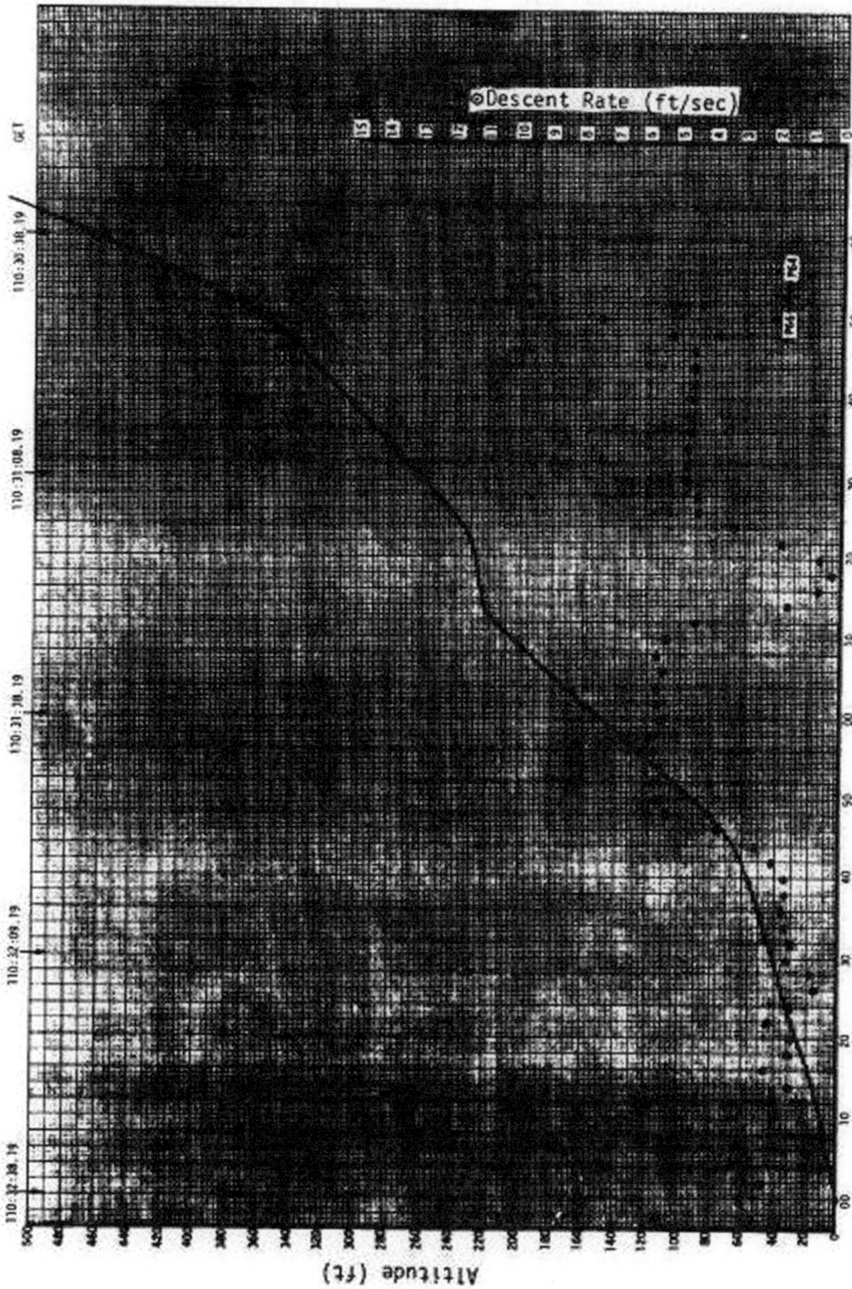

Fig. 10.11 Altitude versus time: Low gate to landing [4]

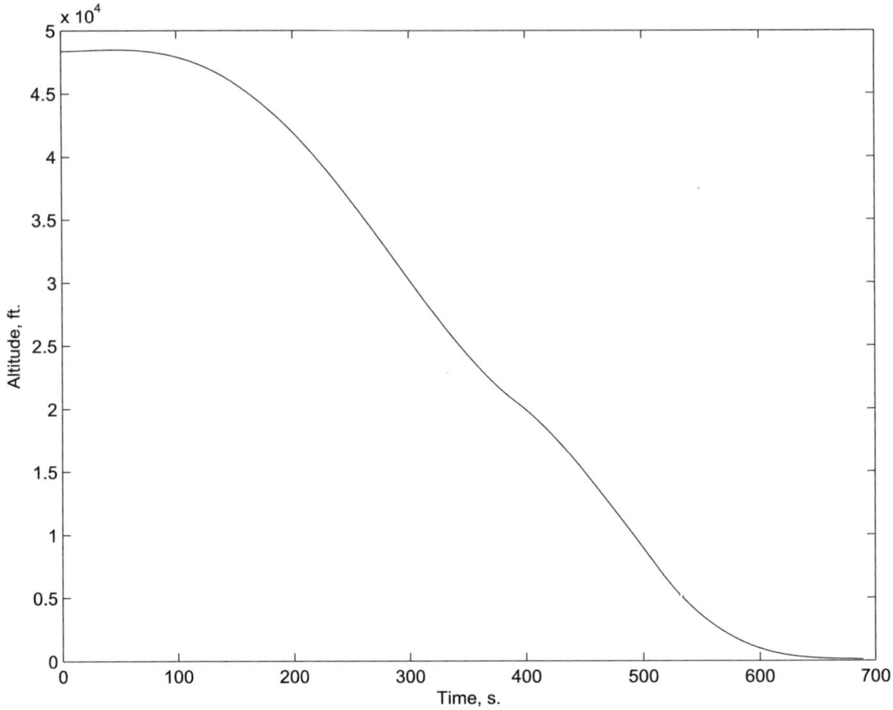

Fig. 10.12 BP and AP: altitude versus time (compare with Fig. 10.11)

10.4 Conclusions

In this Part 1 of the monograph, the Apollo ground-based targeting and on-board ignition and guidance algorithms for the powered lunar descent and landing have been described and reconstructed. Specific simulations have been conducted to compare the results with Apollo 11 and Apollo 12 missions' data. Unlike the existing and published algorithms in the open literature and technical reports, the iterative procedures of the BP-targeting algorithm, and many other procedures of the targeting, ignition and guidance programs have been presented with the explicit content as much as possible. Some minor aspects of the targeting algorithms, such as the specific values of some coefficients are still remaining unknown.

The Apollo ground targeting and two on-board algorithms for the lunar landing maneuver are described with relevant details. The targeting program includes necessary derivations, AP- and BP- targeting algorithms together with constraints, inputs and outputs. Regrading the guidance, the P63 ignition, P63-64 and P66 guidance algorithms have also been described with givens, inputs, constraints, outputs and relevant block-diagrams. The reconstructed targeting programs, properly integrated with guidance programs, provide powered descent from a given parking orbit and landing at any site on the Moon. The targeting programs have been validated by achieving the close agreements in comparing the results of the simulations with the published Apollo post-flight trajectories and guidance performances. In particular, the commanded and actual thrust, and attitude profiles, the altitude, current range to landing site, velocity vector magnitude have been compared to those of the Apollo 11 mission. The values of mass, thrust, flight path angle and central angle at some points of the trajectory are also used in the comparison analysis for Apollo 12. It should be noted that an understanding and proper interpretation of all the algorithms are critically important in constructing the entire flight program to provide a safe and soft landing on the Moon, and on any celestial body in general.

References

1. Apollo Mission 11. Trajectory Reconstruction and Post-flight Analysis. Vol. 1. Project Apollo. TASK MSC/TRW A-50. NAS 9-8166. Houston, Texas. 16 March, 1970, p. 142
2. Azimov DM, Bishop RH (2008) Reconstruction of Apollo Targeting and Guidance for Powered Lunar Descent. New Trends in Astrodynamics and Applications V, Milan, Italy
3. *Project Apollo Coordinate System Standards*, Apollo Program, SE 008-001-1, NASA Office of Manned Space Flight, 1965, (see http://ntrs.nasa.gov/archive/nasa/casi.ntrs.nasa.gov/19700076120_1970076120.pdf)
4. *Apollo 12 Mission Report. Trajectory Reconstruction and Post-flight Analysis*, Vol. 1, 113 p, NASA-TM-X-70272, Houston, Texas, 1970
5. Klumpp AR (1971) "Apollo Lunar-Descent Guidance," Apollo Guidance Navigation Control, R-695. Massachusetts Institute of Technology, MIT Instrumentation Laboratory Report, Cambridge, MA
6. *Apollo Experience Report - Mission Planning for Lunar Module Descent and Ascent*, NASA TN D-6846, June, 1972
7. Azimov, D.M., and Bishop, R.H. Apollo Targeting and G uidance for Pin-Point Landing. Vol. 1. Apollo ground targeting and on-board guidance: theory and algorithms. Internal Report: UT-LENT No. 01-08-02. March 24, 2008. 69 p
8. Apollo 11 Lunar Landing Mission. Press Kit. NASA. July 6, 1969. 100 p
9. Yang, T.L., "A Targeting Scheme for Fuel Optimal Rocket Trajectories With Applications to the LM Descent Braking Phase," Technical Memorandum TM - 71-2014-1, Bellcomm, January 22, 1971

Part II
Integrated Apollo-Class Real-Time Targeting and Guidance for Powered Descent

Chapter 11
Extension and Enhancement of Apollo Guidance

11.1 Introductory Remarks

Part 2 of this monograph presents an extension and enhancement of the Apollo ground-based targeting and on-board guidance design by enabling explicit determination of the state and thrust vector control and on-board targeting capabilities. These capabilities are developed utilizing a new closed-form solution for the constant-thrust arc of the braking phase of the powered lunar-descent trajectory replacing the Apollo's manual ground-based iterative approach. The main elements of the Apollo targeting theory and algorithms, including the key properties of the quartic phases on the descent trajectories, are retained. It is demonstrated that the implementation of the proposed thrust arc solution, unlike the Apollo algorithms, allows us to avoid repeated iterations on the lander's state and target state vectors, and facilitates the on-board execution of targeting and re-targeting procedures at any time during the descent, thereby extending and enhancing the original Apollo targeting and guidance design. It is shown that the shape of the BP constant-thrust arc is highly dependent on initial mass and propulsion system parameters. The analytic solution process is explicit in terms of targeting and guidance parameters, while remaining generic with respect to planetary body and descent trajectory design. These features increase the feasibility of extending the proposed integrated targeting and guidance design to future cargo and robotic landing missions. NASA's long-term goals for future cargo and robotic landing missions to the planets and asteroids suggest the necessity of developing new concepts and methods of autonomous on-board guidance for planetary descent yielding safe and precise landings.

© The Author(s), under exclusive license to Springer Nature Switzerland AG 2026

D. M. Azimov and R. H. Bishop, *Apollo-Class Targeting and Guidance for Lunar Powered Descent and Precision Landing*, Springer Aerospace Technology,
https://doi.org/10.1007/978-3-031-91088-3_11

11.2 Enriching Lunar-Descent Guidance and Targeting with New Capabilities

The successful Apollo missions demonstrated that on-board guidance coupled with ground-based targeting was a viable option for safe landings on the lunar surface [1]. The Apollo-era ground-based targeting and on-board guidance procedures, and many other programs developed since that time, use iterative processes and approximations to determine the target states, state and control parameters [2]. Recently, the results describing the application of analytic methods to planetary landing have been reported [3, 4]. These analytic approaches generally do not use the Apollo-class designs as a starting point. Our approach has been to extend and enhance the proven Apollo targeting and guidance design to allow for a real-time on-board implementation of a proposed design [5, 6]. Creating an on-board targeting capability will impact the robustness of the landing precision using the Apollo design of specifying BP- and AP-quartic targets, but doing so during the descent. The potential benefits of an on-board targeting capability include the ability to re-designate the landing site autonomously at any time, the ability to perform a pinpoint landing, and the ability to utilize on-board terrain mapping for the landing site re-designation to avoid hazardous sites [2]. These capabilities are important, in particular, from the point of view of future cargo and robotic landing missions. Reconstruction, qualitative extension and enhancement of original Apollo-era lunar descent guidance programs can serve as a starting point of the development of new concepts and methods for autonomous, on-board guidance to provide safe and precise landings [7].

11.3 Apollo Descent and Landing Maneuver Trajectory

As briefly described in Chap. 2, a typical Apollo descent and landing maneuver began with transferring the lander from an initial circular lunar orbit to an elliptical entry orbit at its apolune [8]. The lander initiated the ullage followed by the trim phase with minimum thrust. After approximately 26 s, near or at the perilune of the entry orbit, the lander used the DPS to initiate the braking phase (BP) with nearly constant thrust (typically, 93.8–94.8% of maximum allowable thrust) which was not the same as the commanded thrust. At the throttle control recovery (TCR) point, the commanded thrust level was decreased to 57% and the commanded and actual thrusts coincided. The trajectory then transitioned to the second part of the BP which lasted 120 s (this time was specified). This part of the BP was approximated by the BP-quartic. The BP ended at the start of the approach phase which lasted up to 160 s. This phase was also approximated by the AP-quartic. The approach phase ended at the initial point of the TDP at about 30 m altitude and 11 m ground range to the landing site and at about 30 s to landing. The TDP was nearly vertical and a typical descent and landing trajectory was nearly planar. The AP-quartic was described in a closed-form, while the BP-quartic required iterations to satisfy the constraints and the boundary conditions as the state at initial point of this quartic was not computed analytically. Here the Apollo-class means that the targeting and guidance capabilities are based on the main concepts of the original Apollo guidance and targeting routines and their integrated execution [9].

11.4 Previous Studies

As mentioned in Chap. 3, the concept of guidance as a continuation of an optimal control problem was first described by Kelley in 1961 [10]. Klumpp adjusted and modified Cherry's results to obtain a guidance law applicable to powered descent by employing the quartic polynomials (1971) and iterative procedures for BP targeting [8, 11]. The main characteristic of this guidance law was that it used the target states at points BFT and APT computed before the mission (see Fig. 11.1).

As Klumpp notes in Ref. [8], the first significant part of the braking phase before the TCR point had no closed-form solution. Consequently, the requirements for ground-based targeting simulations to numerically synthesize the descent trajectory have been determined. Any on-board implementation of the targeting scheme would

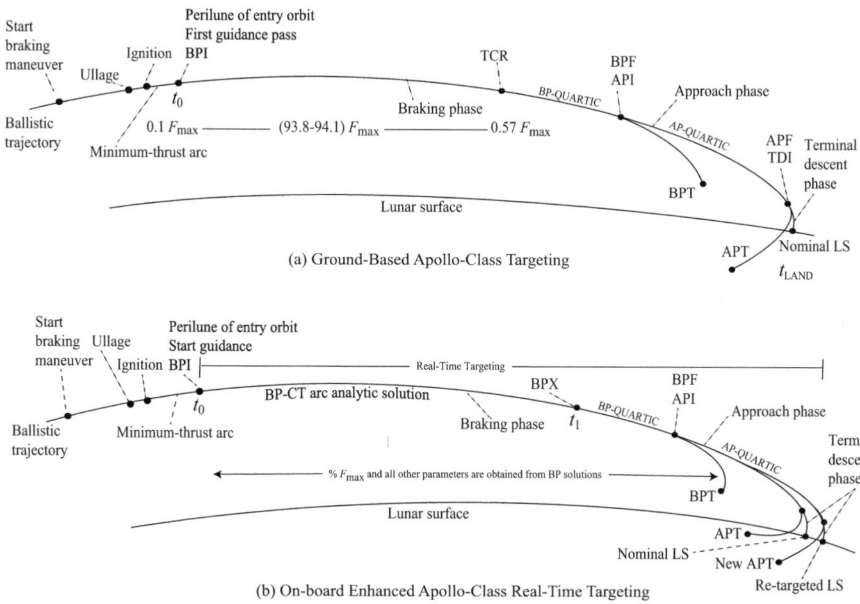

Fig. 11.1 Original Apollo descent trajectory design (with ground-based targeting) and its development (proposed extension and enhancement). BPI and API mean the BP- and AP-initial points, BPX is the final point of BP-CT arc, APF is the AP-final point, TDI is the initial point of TDP, and BPT and APT are the BP- and AP-target points [9]

have led to uncertainties in the number of iterations in realtime (even if the computational resources had been available) and potential convergence issues in computing the BP- and AP-target states.

There exists a number of classes of the analytical solutions developed for the descent guidance for uniform, linear central, and central Newtonian fields in the context of the trajectory optimization problem [6–13]. The uniform field solutions representing the "bilinear tangent steering" law and "sine-cosine steering" law that were developed by Yang [12]. Both laws were shown to be *nearly optimal*, but they do not provide the targeting solutions. R. Jaggers developed the Shuttle Powered Explicit Guidance (PEG) equations (1974) (JSC contract) [14, 15]. These equations were based on the linearization of the Newtonian field (developed by D. Jezewski (JSC)) and by using cubic polynomial for the position vector (Linear Tangent Guidance) [16]. These approximations do not provide a solution to the targeting or guidance problem in lunar landing because of the instability of the state vector components. The state vector becomes unstable after about 40–50 sec. of thrusting (at nearly 15 km altitude) as this vector is described in special (table) functions, such as integral sine and integral cosine. The linear approximation works only if the thrust angle is very small and $r - r_0 < 7\,\text{km}$ which is not the case of the descent maneuver [17]. Consequently, the corresponding solutions do not satisfy the lunar descent requirements. The general approach to the reconstruction of the guidance performance and to the description of the Apollo BP by closed- form thrust arc solutions of a trajectory optimization problem have been reported by the authors [5, 6]. Application of the closed-form solutions developed specifically for this phase allows us to enable the targeting capability [6]. In these solutions, the thrust vector is aligned opposite to the velocity vector during the flight in the Newtonian field. It can be shown that directing the thrust along the velocity vector allows one to obtain the approximate analytic solutions for a "gravity turn" for a spacecraft launched vertically for an orbit insertion maneuver [18]. If a solution is explicit, relatively simple, highly accurate and requires no repeated iterative procedures, then it may be a good candidate to support the development of targeting and guidance laws for on-board implementation. This part of the monograph aims to achieve this main objective.

11.5 Extension and Enhancement of Guidance and Targeting

We aim to extend and enhance an Apollo-class on-board guidance and ground-based targeting by replacing it by the development of integrated on-board targeting and guidance design. It will be demonstrated that a lunar-descent trajectory can be analytically synthesized and designed by considering the first part of the BP as a CT arc. The proposed solutions allow us to execute the BP- and AP-targeting routines once and facilitate retargeting operations (see Fig. 11.2). Therefore, the implementation of the explicit closed-form solutions in the design of powered descent and

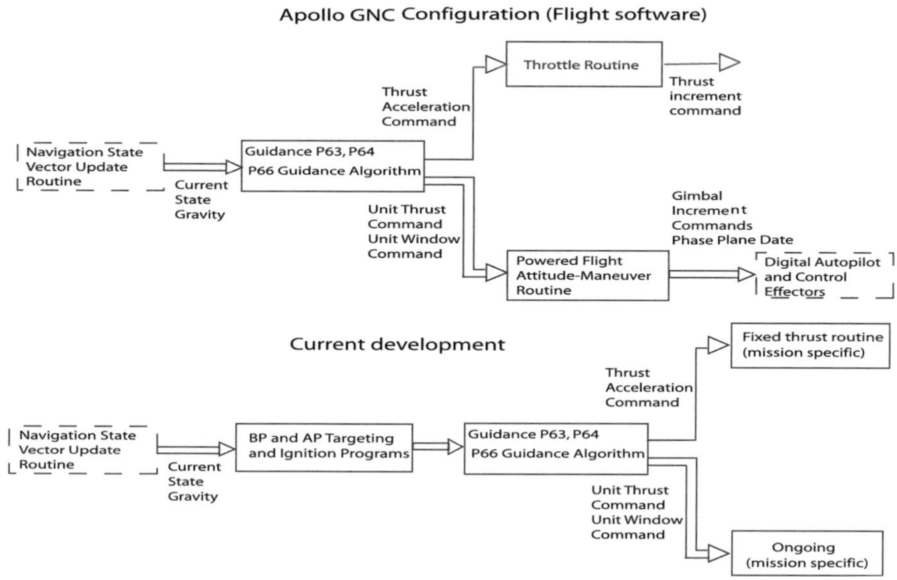

Fig. 11.2 Original form and proposed modification of Apollo GNC configuration

landing maneuvers would represent a qualitative enhancement of the Apollo targeting and guidance programs. In general, the proposed design is applicable to cargo, robotic and transportation landing missions to the Moon, Mars, other planets, planetary moons and asteroids [2, 6]. The terminology used here is based on the general definitions, conventions and the flowchart diagrams employed in the Apollo missions [1, 8, 19, 20]. For the comparison and verification purposes, the Apollo 12 post-flight trajectory data and the optimal reference trajectory data have been analyzed along with the proposed CT solutions. A powered descent trajectory has been designed and on-board targeting and retargeting capabilities including the landing accuracy have been discussed to illustrate the proposed extension and enhancement of the Apollo guidance and targeting.

References

1. Apollo 12 Mission Report (1970) Trajectory reconstruction and post-flight analysis, vol 1. NASA-TM-X-70272, Houston, Texas, p 113
2. Paschall S, Brady T, Fill T, Sostaric R (2009) Lunar landing trajectory design for onboard hazard detection and avoidance. In: 32nd annual AAS guidance and control conference, Breckenridge, CO
3. Chomel CT, Bishop RH (2009) An analytical lunar descent guidance algorithm. AIAA J Guid Control Dyn 32(3):915–926
4. Sostaric R (2007) Powered descent trajectory guidance and some considerations for human lunar landing. In: AAS guidance and control conference, AAS 07–051, Breckenridge, CO

5. Azimov DM, Bishop RH (2008) Reconstruction of Apollo targeting and guidance for powered lunar descent. In: New trends in astrodynamics and applications V, Milan, Italy
6. Bishop RH, Azimov DM (2008) Improved Apollo targeting for onboard lunar landing site re-designation. In: New trends in astrodynamics and applications V, Milan, Italy
7. Epp CD, Smith TB (2008) The autonomous precision landing and hazard detection and avoidance technology (ALHAT). In: Go for lunar landing: from terminal descent to touchdown conference, panel 4: GNC. http://ser.sese.asu.edu/GO/
8. Klumpp AR (1971) Apollo lunar-descent guidance, Apollo guidance navigation control, R-695. Massachusetts Institute of Technology, MIT Instrumentation Laboratory Report, Cambridge, MA
9. Azimov DM, Bishop RH (2018) Integrated targeting and guidance for powered planetary descent. J Astronaut Sci 65:229–259. https://doi.org/10.1007/s40295-017-0126-4
10. Kelley HJ (1962) Guidance theory and extremal fields. IEEE Trans Autom Control 5(2):75–82
11. Klumpp AR (1968) Manually retargeted automatic landing system for the lunar module (LM). AIAA J Spacecr Rocket 5(2):129–138
12. Yang TL (1971) A targeting scheme for fuel optimal rocket trajectories with applications to the LM descent braking phase. Technical memorandum TM—71-2014-1, Bellcomm
13. Azimov DM (2010) Extremal analytical solutions for intermediate-thrust arcs in the Newtonian field. AIAA J Guid Control Dyn 33(5):1550–1565
14. Jaggers RF (1974) Asymmetrical booster ascent guidance and control system design study, vol 5. In: Space shuttle powered explicit guidance. Final report. Contract NAS9-13568, p 41
15. Jaggers RF (1976) Proposed powered explicit guidance thrust integrals derivation/implementation. Design note no. 1.4-7-38. McDonnell Douglas Technical Services Co. Houston Astronautics Division, p 50
16. Jezewski DJ (1972) Optimal analytic multiburn trajectories. AIAA J 10(5):680–685
17. Azimov DM, Bishop RH (2001) Extremal rocket motion with maximum thrust in a linear central field. J Spacecr Rocket 38(5):765–776
18. Boltz FW (1989) Analytic solution for vertical launch into orbit. J Astronaut Sci 37:491–511
19. Cherry GW (1964) E-guidance—a general explicit, optimizing guidance law for rocket-propelled spacecraft, R-456. Massachusetts Institute of Technology, MIT Instrumentation Laboratory Report, Cambridge, MA
20. Klumpp AR (1974) Apollo lunar descent guidance. Automatica 10. Paragon Press, pp 133–146

Chapter 12
Optimal Control Problem and Extremals

Assume that the lander DPS is characterized by the given constant I_{sp} and F_{max}. Then the exhaust velocity, c and the maximum mass-flow rate, β_{max} are

$$c = I_{sp} g_0, \ \beta_{max} = F_{max}/c. \tag{12.1}$$

Similar to the Apollo DPS constraints, it is assumed that the maximum allowable thrust is not achievable on the descent trajectory. A constant exhaust velocity implies that the mass-flow rate satisfies the condition

$$\beta \leq \beta_{max}. \tag{12.2}$$

12.1 Problem Statement

Consider the planar motion of the lander center of mass in a Newtonian field. If a polar coordinate system $Or\theta$ with the origin at the Moon center of gravitational attraction, then the spacecraft motion can be described by the differential equations (see Fig. 12.1) [1, 2]:

$$\dot{u} = \frac{c\beta}{m} e_r - \frac{\mu}{r^2} + \frac{w^2}{r}, \tag{12.3}$$

$$\dot{w} = \frac{c\beta}{m} e_\theta - \frac{uw}{r}, \tag{12.4}$$

$$\dot{r} = u, \tag{12.5}$$

$$\dot{\theta} = \frac{w}{r}, \tag{12.6}$$

$$\dot{m} = -\beta. \tag{12.7}$$

© The Author(s), under exclusive license to Springer Nature Switzerland AG 2026

D. M. Azimov and R. H. Bishop, *Apollo-Class Targeting and Guidance for Lunar Powered Descent and Precision Landing*, Springer Aerospace Technology,
https://doi.org/10.1007/978-3-031-91088-3_12

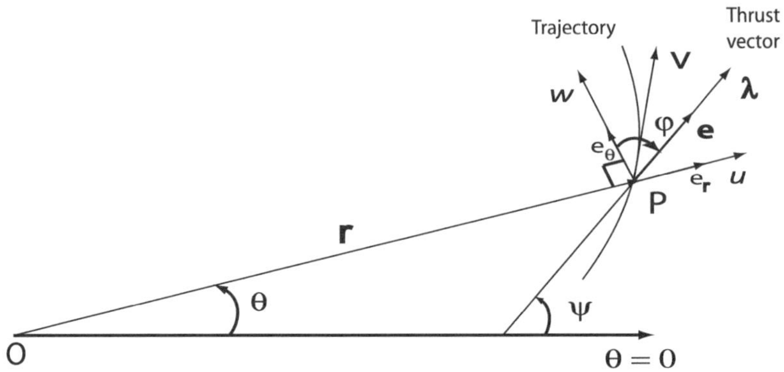

Fig. 12.1 Polar coordinate system defined in the $Xp - Zp$ plane and illustration of angles [3]

The variables r, θ, u, w and m are the state variables and assumed to be real continuous functions of time, although their derivatives may have discontinuities. It is assumed that all variables in Eqs. (12.3)–(12.7) are defined on the interval $[t_0, t_1]$. The initial time of flight, denoted by t_0, is fixed, but the final time of motion, denoted by t_1, may be fixed or free. The variables e_r, e_θ and β are the control variables subject to the constraints

$$e_r^2 + e_\theta^2 = 1, \ 0 \leq \beta \leq p_1 \beta_{max}, \tag{12.8}$$

where p_1 $(0 < p_1 \leq 1)$ is the constant coefficient of the thrust percentage to be determined. The constants c and βmax are known and computed using Eq. (12.1). At the given initial time, t_0, the initial state of the lander is defined by the perilune of the entry orbit with known perilune and apolune altitudes, h_p and h_a, respectively, by the orbital velocity at this point, and by the initial mass, as

$$r(t_0) = r_0, \ \theta(t_0) = \theta_0, \ u(t_0) = u_0 = 0, \ w(t_0) = w_0, \ m(t_0) = m_0. \tag{12.9}$$

At the final time, t_1, the final state is defined at the TCR point by

$$r(t_1) = r_1, \ v(t_1) = v_1, \tag{12.10}$$

where $v = \sqrt{u^2 + w^2}$ and t_1 are not specified. The functional J of the problem is defined as

$$J = m_0 - m_1, \tag{12.11}$$

where $m_1 = m(t_1)$. The functional J and its high-order derivatives are considered to be continuous on $[t_0, t_1]$. Note that, in general, J can be an implicit or explicit function of t_0 and t_1. The problem is to find an optimal control and an optimal trajectory which satisfy Eqs. (12.3)–(12.10) and transfers the lander from the initial state to the final state while minimizing J. Note that if $p_1 = 1$, that is if F_{max} is achievable, then the problem described in Eqs. (12.3)–(12.11) is a typical Mayer's variational or optimal control problem. The thrust percentage on the Apollo pre-quartic BP was 93.8% of F_{max}, that is, $p_1 = 0.938$.

12.2 Extremals of the Problem

It can be shown that in the context of the problem statement given above, the analysis of the first-order necessary conditions of optimality and the Weierstrass condition can lead to the following conclusions [3]:

(1) An optimal trajectory may contain a combination of a null-thrust arc, where $\beta = 0$ and $\chi \leq 0$, an intermediate-thrust arc, where $0 < \beta < p_1\beta max$ and $\chi \equiv 0$, and a constant-thrust arc, where $\beta = p_1\beta max$ and $\chi \geq 0$.
Here

$$\chi = \frac{c}{m}\lambda - \lambda_m \tag{12.12}$$

is the switching function, and λ and λ_m are the primer vector magnitude and the Lagrange multiplier associated with the lander mass, respectively.
(2) On the optimal trajectory, the thrust direction cosines must satisfy the conditions:

$$e_r = \lambda_u/\lambda,\ e_\theta = \lambda_w/\lambda, \tag{12.13}$$

where $\lambda = ||\boldsymbol{\lambda}|| = \surd\lambda_u + \lambda_w$ is the magnitude of the primer vector, $\boldsymbol{\lambda}$. These conclusions allow us to express the relationships in Eqs. (12.3)–(12.7) and the first-order necessary conditions of optimality by a set of canonical equations formed separately for each thrust arc mentioned above and by the transversality conditions

$$\lambda_{s1} = -\nu\frac{\partial J}{\partial x_s},\ \ s = 1, .., l \leq 5, \tag{12.14}$$

where x_s is any state variable not specified at t_1. For a normal solution, without loss of generality, one can use $\nu = 1$ in Eq. (12.14). If the state vector is defines as $\mathbf{x} = \mathbf{x}(u,\ w,\ r,\ \theta,\ m)$, then it can be shown that by taking into account Eq. (12.13), the CT arc can be described by the canonical equations [2]:

$$\dot{\mathbf{x}} = \left[\frac{\partial H}{\partial \boldsymbol{\lambda}}\right]^T, \quad \dot{\boldsymbol{\lambda}} = -\left[\frac{\partial H}{\partial \mathbf{x}}\right]^T,$$

or in a scalar form as

$$\dot{u} = \frac{c\beta}{m}\frac{\lambda_u}{\lambda} - \frac{\mu}{r^2} + \frac{w^2}{r}, \tag{12.15}$$

$$\dot{w} = \frac{c\beta}{m}\frac{\lambda_w}{\lambda} - \frac{uw}{r}, \tag{12.16}$$

$$\dot{r} = u, \tag{12.17}$$

$$\dot{\theta} = \frac{w}{r}, \tag{12.18}$$

$$\dot{m} = -\beta, \tag{12.19}$$

$$\dot{\lambda}_u = \lambda_w \frac{w}{r} - \lambda_r, \tag{12.20}$$

$$\dot{\lambda}_w = -2\lambda_u \frac{w}{r} + \lambda_w \frac{u}{r} - \frac{\lambda_\theta}{r}, \tag{12.21}$$

$$\dot{\lambda}_r = \lambda_u (\frac{w^2}{r^2} - 2\frac{\mu}{r^3}) - \lambda_w \frac{uw}{r^2} + \lambda_\theta \frac{w}{r^2}, \tag{12.22}$$

$$\dot{\lambda}_\theta = 0, \tag{12.23}$$

$$\dot{\lambda}_m = \frac{c\beta}{m^2}\sqrt{\lambda_u^2 + \lambda_w^2}, \tag{12.24}$$

with the Hamiltonian

$$H = \lambda_u \left(\frac{w^2}{r} - \frac{\mu}{r^2}\right) - \lambda_w \frac{uw}{r} + \lambda_r u + \lambda_\theta \frac{w}{r} + \left(\frac{c}{m}\lambda - \lambda_m\right)\beta, \tag{12.25}$$

where (λ_u, λ_w) and (u, w) are the components of λ and v, respectively, and λ_r and λ_θ are the Lagrange multipliers related to r and θ respectively, and β is determined using the sign of χ. A trajectory satisfying the above-given canonical system, Eqs. (12.15)–(12.24), is said to be an extremal of the problem and an optimal trajectory with minimum J is one of the members of the family of extremals that also satisfy Eqs. (12.9), (12.10) and (12.14). Due to the fact that the first part of the BP in Apollo was flown with a nearly constant thrust, and the thrust on an intermediate-thrust arc is known to be mainly variable, we will consider only the extremal solutions for the CT arcs with $\beta = p_1\beta_{max}$, $p_1 = const < 1$ and which satisfy Eqs. (12.8)–(12.10), (12.14), and Eqs. (12.15)–(12.24) with the hamiltonian H. It should be noted here that for various maneuvers, the extremal trajectories which satisfy the boundary and transversality conditions have been shown to be very close to optimal trajectories [4]. For the CT arcs, the following integrals of Eqs. (12.15)–(12.24) are known [2]:

$$H = C, \ \lambda_\theta = \lambda_{\theta 0}, \ m = m_0 - \beta t, \tag{12.26}$$

where C, $\lambda_{\theta 0}$ and m_0 are the integration constants. The first expression in Eq. (12.26) is known as the Hamiltonian integral which results from an implicit independence of H on time, t. Similarly, the integral for λ_θ is known as the cyclic integral as H does not explicitly depend on θ.

References

1. Azimov DM, Bishop RH (2018) Integrated targeting and guidance for powered planetary descent. J Astronaut Sci 65:229–259. https://doi.org/10.1007/s40295-017-0126-4
2. Azimov DM, Bishop RH (2005) Optimal trajectories for space guidance. In: Annals of New York academy of sciences, vol 1065. New York, NY, pp 1–21
3. Lawden DF (1963) Optimal trajectories for space navigation. Butterworths, London
4. Azimov DM (2010) Extremal analytical solutions for intermediate-thrust arcs in the Newtonian field. AIAA J Guid Control Dyn 33(5):1550–1565

Chapter 13
Extension of Braking-Phase Targeting

13.1 Braking-Phase Quartic

The braking phase (BP) is comprised of two subarcs. The first subarc, known as the pre-quartic BP, starts at the perilune of the entry orbit and ends at the TCR point. The second subarc, represented as a quartic and known as the BP-quartic, begins at the TCR point and ends at the initial point of the AP. Connecting the pre-quartic BP with the entry orbit on the one side and with the BP-quartic on the other side was accomplished in the Apollo design using ground-based simulations in an iterative fashion. As stated before, our objective is to represent the pre-quartic BP analytically and to use continuity conditions to fit the pre-quartic BP into the descent trajectory. The current position, velocity, acceleration, jerk and snap vectors on the BP- quartic can be determined by the following polynomial formulas [1]:

$$\mathbf{r} = \mathbf{r}_T + T\mathbf{v}_T + \frac{T^2}{2}\mathbf{a}_T + \frac{T^3}{6}\mathbf{j}_T + \frac{T^4}{24}\mathbf{s}_T, \tag{13.1}$$

$$\mathbf{v} = \mathbf{v}_T + T\mathbf{a}_T + \frac{T^2}{2}\mathbf{j}_T + \frac{T^3}{6}\mathbf{s}_T, \tag{13.2}$$

$$\mathbf{a} = \mathbf{a}_T + T\mathbf{j}_T + \frac{T^2}{2}\mathbf{s}_T, \tag{13.3}$$

$$\mathbf{j} = \mathbf{j}_T + T\mathbf{s}_T, \tag{13.4}$$

$$\mathbf{s} = \mathbf{s}_T, \tag{13.5}$$

where the target referenced time, T which is computed backwards from the target point to the point of interest on the quartic. The subscript T indicates that the corresponding parameter is computed or given at the target point. Note, that $T = 0$ at the target point, which means that $T < 0$ during the maneuver. But the lander never reaches the target point of the quartic, which means that $T \neq 0$. According to the Apollo guidance design strategy, the nominal descent trajectories on the second part

© The Author(s), under exclusive license to Springer Nature Switzerland AG 2026

D. M. Azimov and R. H. Bishop, *Apollo-Class Targeting and Guidance for Lunar Powered Descent and Precision Landing*, Springer Aerospace Technology,
https://doi.org/10.1007/978-3-031-91088-3_13

of the BP and on the AP are represented by quartics lying in the $X_p - Z_p$ plane of the platform coordinate system [2] (see Fig. 4.3 of Chap. 4). This means that the Y-components of all vectors on the nominal trajectory are zero. If we knew the target states, $\mathbf{r}_T$, $\mathbf{v}_T$, $\mathbf{a}_T$, $\mathbf{j}_T$, $\mathbf{s}_T$ and T, we could use Eq. (13.3) to compute the state at TCR. The challenge is to find the BP-quartic target state that leads to a state at TCR that connects through backward propagation in a Newtonian gravity field, and then through the pre-quartic BP constant-thrust arc to the perilune of the entry orbit. In the Apollo design, some components of the BP-quartic target state were specified (working backwards from the desired landing site through the TDP and AP-quartic), while the remaining components others were determined numerically using simulations. In our approach, we intend to analytically compute the remaining BP-quartic components to smoothly connect the descent trajectory. Each of the iterations in the BP-targeting routine requires the execution of the AP-targeting routine in order to match the terminal conditions of the BP and the initial conditions of the AP, and to find the right conditions at the TCR point. Note that (a) the number of iterations is unknown, and their convergence is not guaranteed in advance for arbitrary final conditions for both AP and BP, and (b) the actual thrust is not the same as the commanded thrust until the TCR point is reached. These two factors illustrate the importance of describing the entire BP by a closed-form solution to avoid the iterations and repeated runs of the AP-targeting routine. This demonstrates the utility of the development of the proposed CT arc solutions in guidance and targeting. It will be shown that such solutions enable real-time executions of the BP- and AP-targeting algorithms only once.

13.2 Closed-Form Solutions for Braking Phase

It is known that if a spacecraft escapes the Newtonian gravitational field, then maximization of its total energy requires that at the final time of this maneuver the thrust vector is aligned with the spacecraft velocity vector [3]. Similarly, if it is required to minimize the total energy, then from the transversality condition it can be shown that at the final time the thrust vector is aligned opposite to the velocity vector. As zero kinetic energy implies zero velocity, the process of decreasing both altitude and velocity, and reaching zero total velocity at landing is similar to minimizing the kinetic energy at landing. From this point of view it is reasonable to consider aligning the thrust vector opposite to velocity vector for descent and landing purposes. In particular, when non-zero final velocity is desired, such an alignment of the thrust vector can be used on the portion of CT arc of the descent trajectory. It will be shown that this approach to determine the thrust direction enables the integration of Eqs. (12.15)–(12.24) in elementary functions, except for the polar angle which is expressed in quadratures. The corresponding solutions for the CT arc are then used to describe the pre-quartic BP arc. For further analysis of the descent trajectory, it is convenient to consider the polar coordinate system defined in the $X_p - Z_p$ plane of the platform coordinate system because the Apollo nominal descent trajectories were considered in this plane [1]. The coordinate systems used in the studies are shown in Fig. 4.3 of Chap. 4. Based on the assumptions about the thrust vector alignment, consider the expression

$$\boldsymbol{\lambda} = -q(t)\mathbf{v}, \tag{13.6}$$

where the primer vector, $\boldsymbol{\lambda}$, determines the thrust direction, $q(t)$ is an unknown function, continuous and differentiable with respect to time. It will be shown that $q(t) > 0$ and $\dot{q}(t) > 0$ at any time on the CT arc that satisfies the boundary conditions in Eqs. (12.9) and (12.10). The scalar form of Eq. (13.6) is written as

$$\lambda_u = -q(t)u \qquad \lambda_w = -q(t)w. \tag{13.7}$$

Taking into account that $\lambda_u = \lambda \sin\varphi$ and $\lambda_w = \lambda \cos\varphi$, where φ is the thrust angle, defined as the angle between ˘ and perpendicular to **r**, from Eq. (13.7) it follows that

$$u = w \tan\varphi. \tag{13.8}$$

Equation (13.8) shows that the thrust direction is collinear with the velocity direction (see Eq. (13.6)). If ψ is the angle between the thrust direction and the reference axis is $\theta = 0$, then [3]

$$\psi = \theta - \varphi + \frac{\pi}{2}. \tag{13.9}$$

Taking the time derivative of Eq. (13.8), and substituting $\dot{u}$ and $\dot{w}$ from Eqs. (12.15) and (12.16) yields

$$rw^2 - r^2 w\dot{\varphi} - \mu\cos^2\varphi = 0. \tag{13.10}$$

Considering the derivative of Eq. (13.9) and substituting into Eq. (13.10) yields

$$w\dot{\psi} = \frac{\mu}{r^2}\cos^2\varphi. \tag{13.11}$$

Now, consider the derivative $\dot{\lambda}_w$, obtained from Eq. (13.7) as

$$\dot{\lambda}_w = -w\dot{q} - \dot{w}q. \tag{13.12}$$

Substituting the right hand sides of (12.21) for $\dot{w}$ into Eq. (13.12) yields

$$\dot{q}rw + qr\frac{c\beta}{m}cos\varphi = \lambda_\theta. \tag{13.13}$$

Note that on the CT arc we have $m = m_0 - \beta t$, and also Eq. (12.23) for $\dot{\lambda}_\theta$ ireveals a new integration constant,

$$\lambda_\theta = \lambda_{\theta 0}. \tag{13.14}$$

If the performance index does not depend on the polar angle and the final polar angle is not specified, then the transversality condition for λ_θ yields $\lambda_\theta = \lambda_{\theta 0} = \frac{\partial J}{\partial \theta_1} = 0$. In this case, Eq. (13.13) is rewritten as

$$w = -\frac{q}{\dot{q}}\frac{c\beta}{m}\cos\varphi,\ q > 0,\ \dot{q} \neq 0. \tag{13.15}$$

It can be shown that although the case with $\dot{q} = 0$ simplifies the solution process considerably, the resulting solutions are not applicable to the descent maneuver. Employing Eqs. (13.8) and (13.15), one can show that the magnitude v of the velocity vector $\mathbf{v}$ can be found in the form:

$$v = \frac{q}{\dot{q}}\frac{c\beta}{m}, \tag{13.16}$$

which shows that $\dot{q} > 0$. Then, substitution of the right hand side of the formula

$$\frac{c\beta}{m}cos\varphi = \dot{w} + w\frac{\dot{r}}{r}, \tag{13.17}$$

obtained from the equation for $\dot{w}$ in Eq. (12.16) into Eq. (13.13) yields

$$rw\dot{q} + rq\dot{w} + wq\dot{r} = \lambda_{\theta 0}, \tag{13.18}$$

which can be integrated to yield another integral of the canonical system, Eqs. (12.15)–(12.24) as

$$rqw = \lambda_{\theta 0}t + a_1, \tag{13.19}$$

where a_1 is an integration constant. Due to the transversality condition, $\lambda_\theta = \lambda_{\theta 0} = 0$, Eq. (13.19) reduces to

$$rqw = a_1. \tag{13.20}$$

One more equation for φ can be obtained by using Eq. (13.8) and rewriting the differential equations for u and w in Eqs. (12.15) and (12.16) in the form

$$v\dot{\varphi} = \left(\frac{\mu}{r^2} - \frac{v^2}{r}\right)cos\varphi, \quad \dot{v} = \frac{\mu}{r^2}sin\varphi - \frac{c\beta}{m}. \tag{13.21}$$

Differentiating Eq. (13.16) with respect to time and equating to $\dot{v}$ in Eq. (13.21) yields the following cubic equation in terms of φ

$$\sin^3\varphi - \sin\varphi + \frac{a_1^2}{\mu}\frac{\dot{q}}{q^3}\frac{m}{c\beta}\left(\frac{\beta}{m} + 2\frac{\dot{q}}{q} - \frac{\ddot{q}}{\dot{q}}\right) = 0 \tag{13.22}$$

Note that Eq. (13.22) can also be derived using Eqs. (13.20) and (13.21) and $\dot{r} = -v \sin\varphi = -dr/d\varphi\dot{\varphi}$. The root of interest in Eq. (13.22) is

$$\sin\varphi = \frac{2}{\sqrt{3}}\cos(\eta/3), \tag{13.23}$$

where

$$\cos\eta = -\frac{\sqrt{27}}{2}\frac{a_1^2}{\mu}\frac{\dot{q}}{q^3}\frac{m}{c\beta}\left(\frac{\beta}{m} + 2\frac{\dot{q}}{q} - \frac{\ddot{q}}{\dot{q}}\right). \tag{13.24}$$

At this point, one can search for the magnitude v of velocity vector in the form

$$v = v_0\frac{(1-ht)}{(1+kt)}, \tag{13.25}$$

where v_0, k and h are the arbitrary constants introduced to satisfy the boundary conditions. The justification for searching for the magnitude of velocity vector in the form Eq. (13.25) is that, for $h > 0$ and $k > 0$, it provides a decreasing function of time, allows us to satisfy Eqs. (12.15)–(12.24) and the boundary conditions, introduces two constants for this system, and allows us to integrate Eq. (13.26) in a closed-form. Consequently, the function q can be determined using Eq. (13.16) as a function that satisfies

$$\frac{\dot{q}}{q} = \frac{c}{v_0}\frac{\beta}{m_0}\frac{1+kt}{(1-ht)(1-\frac{\beta}{m_0}t)}. \tag{13.26}$$

Equation (13.26) can be integrated to obtain the new integral

$$q = a_2\frac{(1-ht)^{k_1}}{(1+at)^{k_2}}, \tag{13.27}$$

where a_2 is a constant of integration, and

$$\alpha = \frac{c}{v_0},\ a = \frac{\beta}{m_0},\ k_1 = \frac{\alpha a(h+k)}{h(a-h)},\ k_2 = \frac{\alpha(k+a)}{(a-h)}. \tag{13.28}$$

Note that Eq. (13.16) can be considered as a differential equation with respect to q. As this function is an arbitrary function, one can search for its form using Eq. (13.27). Substitution of this form into Eq. (13.16) yields Eq. (13.25), which provides a positive decreasing function of time with a sufficient number of constants to satisfy the boundary and transversality conditions. The solution for the polar angle, θ can be obtained as follows. From Eq. (13.20) it follows that $w = a_1/(rq)$. Using this expression, the equation for $\dot{\theta}$ in Eq. (12.18) can be rewritten as

$$\dot{\theta} = \frac{w}{r} = \frac{a_1}{qr^2}. \tag{13.29}$$

Similarly, from Eqs. (13.20) and (13.15) one can find that

$$r = \frac{a_1}{qw} = -\frac{\dot{q}}{q^2}\frac{m}{c\beta}\frac{a_1}{\cos_\varphi},$$

or

$$r = -\frac{a_1}{qv\cos\varphi}, \tag{13.30}$$

Also, using Eqs. (13.30), (13.29) can be rewritten in the form

$$\dot{\theta} = \frac{a_1}{qr^2} = \frac{q^3}{a_1\dot{q}^2}\frac{c^2\beta^2}{m^2}\cos^2\varphi. \tag{13.31}$$

Using Eqs. (13.16) and (13.23) allows us to express θ in the form of the quadrature

$$\theta = \frac{1}{a_1}\int_{t_0}^{t_1} qv^2\left(1 - \frac{4}{3}\cos^2(\frac{\eta}{3})\right) dt + \theta_0. \tag{13.32}$$

where θ_0 is the integration constant. Using the expressions for $\cos\varphi$ and $\sin\varphi$ found from Eqs. (13.30) and (13.23), and utilizing the identity $1 - \cos 2\varphi = \sin 2\varphi$ yields

$$1 - \frac{a_1^2}{r^2v^2q^2} = \frac{r^4v^2}{\mu^2}\left(\frac{p_1\beta_{max}}{m} + 2\frac{\dot{q}}{q} - \frac{\ddot{q}}{\dot{q}}\right)^2. \tag{13.33}$$

Equation (13.33) will be used to determine the thrust percentage on the CT arc. Once the main variables, u, w, r, θ and m of Eqs. (12.15)–(12.24) are determined, the costate variables of this system can be readily obtained by means of elementary functions and quadratures. Indeed, using Eqs. (13.8), (13.15) and (13.16), one can write

$$u = -v\sin\varphi, \qquad w = -v\cos\varphi, \tag{13.34}$$

where v and φ can be computed from Eqs. (13.23) and (13.25). Using Eq. (13.34), the costate variables λ_u and λ_w can be determined from Eq. (13.7). The expression for λ_r can be obtained by integrating Eq. (12.22) for this variable with substitutions from Eqs. (13.7), (13.8), (13.15), (13.16), (13.22) and (13.30). The integral for λ_θ is given in Eq. (13.14). By substituting Eq. (13.7) into the equation for $\dot{\lambda}_m$ in Eq. (12.24), one can readily obtain the quadrature for λ_m. To summarize the solution process, the solutions of the canonical system, Eqs. (12.15)–(12.24) are given in the form of functions in terms of q, v and φ as

$$
\begin{aligned}
u &= -v \sin\varphi,\\
w &= -v \cos\varphi,\\
r &= -\frac{a_1}{qv\cos\varphi},\\
\theta &= \frac{1}{a_1}\int_{t_0}^{t} qv^2\left(1-\frac{4}{3}\cos^2(\frac{\eta}{3})\right)dt+\theta_0,\\
m &= m_0(1-at),\\
\lambda_u &= qv\sin\varphi,\\
\lambda_w &= qv\cos\varphi,\\
\lambda_r &= -2\int_{t_0}^{t} qv^2\left(\frac{\beta}{m_0(1-at)}+2\frac{\dot q}{q}-\frac{\ddot q}{\dot q}\right)dt+\lambda_{r0},\\
\lambda_\theta &= \lambda_{m0},\\
\lambda_m &= \frac{c\beta}{m_0^2}\int_{t_0}^{t}\frac{qv}{(1-at)^2}dt+\lambda_{m0},
\end{aligned}
\tag{13.35}
$$

where

$$
\begin{aligned}
q &= a_2\frac{(1-ht)_1^k}{(1-at)_2^k},\\
\frac{\dot q}{q} &= \frac{c}{v_0}\frac{\beta}{m_0}\frac{1+kt}{(1-ht1-at)},\\
v &= v_0\frac{(1-ht)}{(1+kt)},\\
\sin\varphi &= \frac{2}{\sqrt{3}}\cos(\eta/3),\\
\cos\eta &= -\frac{\sqrt{27}\,a_1^2}{2}\frac{1}{\mu}\frac{1}{q^2v}\left(\frac{\beta}{m_0(1-at)}+2\frac{\dot q}{q}-\frac{\ddot q}{\dot q}\right),\\
k_1 &= \frac{\alpha a}{h}\frac{h+k}{a-h},\\
k_2 &= \alpha\frac{k+a}{a-h},
\end{aligned}
\tag{13.36}
$$

and the complete set of the integration constants of Eq. (13.35) are $a_1,\ a_2,\ k,\ h,\ v_0,\ \theta_0,\ m_0,\ \lambda_{r0},\ \lambda_{\theta0}(=0)$, and λ_{m0}. This completes the derivation of the solutions for the CT arc. Note that these solutions are assumed free of navigation and control errors. Validity of Eq. (13.35) with (13.36) has been verified analytically by substituting them into Eqs. (12.15)–(12.24) and obtaining the identities for each equation of this system. No singularities may occur in Eq. (13.35) with (13.36) as

$$
q>0,\ \dot q>0,\ \varphi=\frac{\pi}{2},\ \frac{3\pi}{2},\dots\ a\neq h,\ a<\frac{1}{t_1},\ k_1\neq 0,\ k_2\neq 0. \tag{13.37}
$$

13.3 Initial and Final Conditions for Braking Phase

The pre-quartic BP will be referred to as the BP-CT arc as it will satisfy the final conditions and constraints of the BP. This arc starts at the perilune of the entry orbit and ends at the TCR point. It can be shown that [4]

$$\varphi_0 = \pi, \ \frac{a_1}{a_2} = -\sqrt{\mu p}, \ v_0 = \sqrt{\frac{\mu}{p}}(1+e), \ h = \alpha a - k, \tag{13.38}$$

where $\varphi_0 = \varphi(t_0)$, and the entry orbit parameters are

$$e = \frac{h_a - h_p}{h_a + h_p + 2r_p}, \ and \ p = \frac{(r_p + h_p)}{(1+e)}. \tag{13.39}$$

Without loss of generality, one can choose the angular distance of perilune of the entry orbit to be zero. This selection of the constants φ_0, a_1/a_2, v_0 and h satisfies the continuity conditions for the state vector at a point that connects the perilune of the entry orbit with the BP-CT arc. If t_1 is the amount of time spent on the BP-CT arc, then the final mass, m_1 on this arc can be computed as

$$m_1 = m_0 - \beta t_1, \qquad \beta = p_1 \beta_{max}. \tag{13.40}$$

Taking into account that $r_1 = r(t_1)$ and $v_1 = v(t_1)$ are given at the TCR point, it can be shown that

$$\begin{aligned} h &= \frac{1 + \alpha a t_1 - \delta}{t_1(1-\delta)}, \\ \cos\varphi_1 &= -\frac{a_2}{\sqrt{\mu p}} r_1 v_1 q_1, \\ q(t_1) &= a_2 \frac{(1 - h t_1)^{k_1}}{(1 - a t_1)^{k_2}}, \\ \delta &= \frac{v_0}{v_1}. \end{aligned} \tag{13.41}$$

The constant a_2 can be found using the transversality condition and the condition that $\chi = 0$ at the end point of the BP-CT arc. Indeed, utilizing Eq. (12.11), one can show that

$$\lambda_{m1} = \lambda_m(t_1) = -\frac{\partial J}{\partial m_1} = 1. \tag{13.42}$$

Equations (12.25) and (13.42) can be used to compute the integration constants λ_{m0} and λ_{r0}. In particular, from the transversality condition it follows that $H(t_1) = C = -\partial J/\partial t_1 = 0$. Also, from $\chi(t_1) = 0$ it follows that $c\lambda(t_1) = m_1 \lambda_{m1}$. By substituting the expression $\lambda(t_1) = q_1 v_1$ into this condition, it follows from the latter that

$$a_2 = \frac{m_1}{cv_1} \frac{(1-kt_1)^{k_2}}{(1-ht_1)^{k_1}} > 0. \tag{13.43}$$

Consequently, it can be seen from Eqs. (13.26) and (13.27) that $q > 0$ and $\dot{q} > 0$. From the second expression in Eq. (13.38) one can find that

$$a_1 = -\sqrt{\mu p}\frac{m_1}{cv_1} \frac{(1-kt_1)^{k_2}}{(1-ht_1)^{k_1}}. \tag{13.44}$$

Given βmax, and r_1 and v_1 at the final point of the CT-arc, the thrust percentage, denoted by p_1, on this arc can be computed as a function of t_1 using the transcendental equation

$$1 - \frac{\mu p}{r_1^2 v_1^2 q_1^2} = \frac{r_1^4 v_1^2}{\mu^2}\left[\left(\frac{p_1\beta_{max}}{m} + 2\frac{\dot{q}}{q} - \frac{\ddot{q}}{\dot{q}}\right)^2\right]|_{t=t_1}. \tag{13.45}$$

Equation (13.45) establishes an important relationship between the final position and velocity vectors, and the thrust percentage and the final time on the CT arc. Note that this equation also allows us to compute t_1 as a function of p_1. Consequently, Eq. (13.45) can be used to determine all extremal values of p_1 and t_1. The extremal values of these variables are those for which can Eq. (13.45) can be solved in real functions. The optimal value of p_1 can be determined from the minimum of the fuel expenditure computed by using the expression $p_1 t_1 \beta max$ among all CT extremals with corresponding p_1 and t_1. Once p_1 is determined, the mass-flow rate on the BP-CT arc is computed using Eq. (13.41). This completes the computation of the integration constants and the unknown parameters of the BP-CT arc. These constants are used to provide continuity of the position and velocity vectors at the TCR point.

13.4 Continuity of Position and Velocity Vectors at TCR Point

The continuity conditions at the TCR point are formed by equating the corresponding components of the final position and velocity vectors of the BP-CT arc defined in the $X_p - Z_p$ plane and the same vectors defined on the BP-quartic (see Fig. 4.3 of Chap. 4). Assume that the position, velocity and thrust acceleration vectors at the BP-final point are given as

$$\mathbf{r}_{bf}(r_{xf}, r_{yf}, r_{zf}),\ \mathbf{v}_{bf}(v_{xf}, v_{yf}, v_{zf}),\ \mathbf{a}_{bf}(a_{xf}, a_{yf}, a_{zf}) = \frac{c\beta}{m_b f}\mathbf{u}_F + \mathbf{g}_F, \tag{13.46}$$

where $\mathbf{u}_F = (\sin\psi_F, 0, cos\psi_F)$ and $\mathbf{g}_F = (g\cos\theta_F, 0, -g\sin\theta_F)$ are the unit thrust and gravity vectors at the point being considered, ψ_F is the angle between the thrust vector and local horizontal and given according to the Apollo targeting constraints, and

$$g = \mu/|\mathbf{r}_{bf}|^2, \ \theta_F = \arctan\frac{r_{xf}}{r_{zf}}. \tag{13.47}$$

The subscripts I, M, F and T refer to the values of the corresponding parameters at the initial, middle, final and target points of the corresponding phases, respectively. Then the continuity conditions at this point can be written in the following form:

$$\begin{aligned}
x = r\sin\theta_1 &= r_{xf} + v_{xf}T + a_{xf}\frac{T^2}{2} + j_{xf}\frac{T^3}{6} + s_{xf}\frac{T^4}{24},\\
z = r\cos\theta_1 &= r_{zf} + v_{zf}T + a_{zf}\frac{T^2}{2} + j_{zf}\frac{T^3}{6} + s_{zf}\frac{T^4}{24},\\
v_x = u\sin\theta_1 + w\cos\theta_1 &= v_{xf} + a_{xf}T + j_{xf}\frac{T^2}{2} + s_{xf}\frac{T^3}{6},,\\
v_z = u\cos\theta_1 + w\sin\theta_1 &= v_{zf} + a_{zf}T + j_{zf}\frac{T^2}{2} + s_{zf}\frac{T^3}{6},
\end{aligned} \tag{13.48}$$

where x, z, v_x, v_z are the components of the position and velocity vectors computed at the final point of the BP-CT arc (the TCR point), θ_1 is the polar angle of this point, and T is the target referenced time spent on the BP-quartic trajectory. The target referenced time, T is assumed to be given (typically, 120 s in Apollo 11 and Apollo 12 missions). The variable T should not be confused with the subscript T. These equations allow us to find the unknowns j_{xf}, j_{zf}, s_{xf}, s_{zf} in terms of θ_1 as

$$\begin{aligned}
j_{xf} &= \frac{24}{T^3}(s_1 - s_3\frac{T}{4}),\\
j_{zf} &= \frac{24}{T^3}(s_2 - s_4\frac{T}{4}),\\
s_{xf} &= \frac{24}{T^4}(s_1 - j_{xf}\frac{T^3}{6}),\\
s_{zf} &= \frac{24}{T^4}(s_2 - j_{zf}\frac{T^3}{6}),
\end{aligned} \tag{13.49}$$

where

$$\begin{aligned}
s_1 &= r\sin\theta_1 - k_1,\\
s_2 &= r\cos\theta_1 - k_2,\\
s_3 &= u\sin\theta_1 + w\cos\theta_1 - k_3,\\
s_4 &= u\cos\theta_1 - w\sin\theta_1 - k_4,\\
k_1 &= r_{xf} + v_{xf}T + a_{xf}\frac{T^2}{2},\\
k_2 &= r_{zf} + v_{zf}T + a_{zf}\frac{T^2}{2},
\end{aligned}$$

$$k_3 = v_{xf} + a_{xf}T,$$
$$k_4 = v_{zf} + a_{zf}T.$$

Consequently, the quartic state at the TCR point can be determined completely by using the position and velocity vectors defined at the same point of the BP-CT arc. The BP-target state can now be defined as the final point of the BP-quartic, where $\mathbf{x}_F = (\mathbf{r}_F, \mathbf{v}_F, \mathbf{a}_F, \mathbf{j}_F, \mathbf{s}_F)$ and where the Y-components of all vectors are zeros, and j_F and s_F are computed in closed-form using Eq. (13.49). Note that in the case of the Apollo descent trajectory design, the jerk and snap vectors for the BP-quartic are computed by repeated iterations involving the AP-quartic parameters.

13.5 Conditions for Vertical Component of Velocity Vector

Note that on AP- and BP- quartics the X component of the position vector is decreased, but the Z component of this vector is increased during the motion along these quartics. Here we consider the position and velocity vectors on BP- quartic. The X component of the position vector is given as

$$\dot{x} = r_{xf} + v_{xf}t + a_{xf}\frac{t^2}{2} + j_{xf}\frac{t^3}{6} + s_{xf}\frac{t^4}{24},$$

where the right hand side includes the X components of the state vector which are computed according to formulas given in the previous subsection, and t is time. Let us rewrite the components of jerk and snap vectors as

$$j_{xf} = g_1 \sin\theta_2 - g_2 \cos\theta_2 + g_3 \tag{13.50}$$

and

$$s_{xf} = g_4 \sin\theta_2 + g_5 \cos\theta_2 - g_6, \tag{13.51}$$

where

$$g_1 = \frac{24}{T^3}(r - v_1\frac{T}{4}),$$
$$g_2 = \frac{6}{T^2}v_2,$$
$$g_3 = -k_1 + k_3\frac{T}{4},$$
$$g_4 = \frac{24}{T^4}(r - g_1\frac{T^3}{6}),$$

$$g_5 = \frac{4}{T} g_2,$$
$$g_6 = \frac{24}{T^4}(k_1 + g_3 \frac{T^3}{6}).$$

In this case, at any given time, t, the corresponding component of the velocity satisfies the condition

$$\dot{x} = v_{xf} + a_{xf} t + j_{xf} \frac{t^2}{2} + s_{xf} \frac{t^3}{6} < 0$$

or

$$\dot{x} = e_1 + e_2 \sin \theta_2 + e_3 \cos \theta_2 < 0, \tag{13.52}$$

where

$$e_1 = v_{xf} + a_{xf} t + g_3 \frac{t^2}{2} - g_6 \frac{t^3}{6},$$
$$e_2 = g_1 \frac{t^2}{2} + g_4 \frac{t^3}{6},$$
$$e_3 = -g_2 \frac{t^2}{2} + g_5 \frac{t^3}{6}.$$

The inequality Eq. (13.52) is rewritten as

$$(e_1^2 + e_3^2) \sin^2 \theta_2 + 2e_1 e_2 \sin \theta_2 + (e_1^2 - e_3^2) > 0, \tag{13.53}$$

which is equivalent to

$$(\theta_2 - \theta_{21})(\theta_2 - \theta_{22}) > 0, \tag{13.54}$$

where

$$\sin \theta_{21} = \frac{-e_1 e_2 + e_3 \sqrt{e_2^2 - e_1^2 + e_3^2}}{e_2^2 + e_3^2},$$

$$\sin \theta_{22} = \frac{-e_1 e_2 - e_3 \sqrt{e_2^2 - e_1^2 + e_3^2}}{e_2^2 + e_3^2}.$$

If $\theta_{21} > \theta_{22}$, then the solution to Eq. (13.54) is given as

$$\theta_2 \in (-\infty; \theta_{22}) \cup (\theta_{21}, +\infty) . \tag{13.55}$$

But if $\theta_{21} < \theta_{22}$, then

$$\theta_2 \in (-\infty; \theta_{21}) \cup (\theta_{22}, +\infty) . \tag{13.56}$$

If $\theta_{21} = \theta_{22}$, then one can have the solution to Eq. (13.54) in the form

$$\theta_2 \in (-\infty; \theta_{21}) \cup (\theta_{21}, +\infty) . \qquad (13.57)$$

13.6 Conditions for Horizontal Component of Velocity Vector

Similar procedure can be performed for the Z-component of the velocity vector. If we rewrite the Z-components of jerk and snap vectors in the form

$$j_{zf} = p_1 \sin\theta_2 + p_2 \cos\theta_2 - p_3 \qquad (13.58)$$

and

$$s_{zf} = p_4 \sin\theta_2 + p_5 \cos\theta_2 + p_6, \qquad (13.59)$$

where

$$\begin{aligned} p_1 &= \frac{6}{T^2} v_2, \\ p_2 &= \frac{24}{T^3}(r - v_1 \frac{T}{4}), \\ p_3 &= \frac{24}{T^3} k_2, \\ p_4 &= -\frac{4}{T} p_1, \\ p_5 &= \frac{24}{T^4}(r - p_2 \frac{T^3}{6}), \\ p_6 &= \frac{24}{T^4}(-k_2 + p_3 \frac{T^3}{6}). \end{aligned}$$

For any given time, t, the rate of the Z component of the position vector satisfies the condition

$$\dot{z} = v_{zf} + a_{zf} t + j_{zf} \frac{t^2}{2} + s_{zf} \frac{t^3}{6} > 0,$$

or

$$\dot{z} = h_1 + h_2 \sin\theta_2 + h_3 \cos\theta_2 > 0, \qquad (13.60)$$

where

$$h_1 = v_{zf} + a_{zf} t - p_3 \frac{t^2}{2} + p_6 \frac{t^3}{6},$$

$$h_2 = p_1 \frac{t^2}{2} + p_4 \frac{t^3}{6},$$
$$h_3 = p_2 \frac{t^2}{2} + p_5 \frac{t^3}{6}.$$

Equation (13.60) can be rewritten as

$$(\theta_2 - \theta_{21})(\theta_2 - \theta_{22}) < 0, \tag{13.61}$$

where

$$\theta_{21} = \frac{-h_1 h_2 + h_3 \sqrt{h_2^2 - h_1^2 + h_3^2}}{h_2^2 + h_3^2},$$

$$\theta_{22} = \frac{-h_1 h_2 - h_3 \sqrt{h_2^2 - h_1^2 + h_3^2}}{h_2^2 + h_3^2}.$$

Then the solution to Eq. (13.61) is given as

$$\theta_2 \in (\theta_{21}, \theta_{22}), \tag{13.62}$$

if $\theta_{21} < \theta_{22}$, and

$$\theta_2 \in (\theta_{22}, \theta_{21}), \tag{13.63}$$

if $\theta_{21} > \theta_{22}$. If $\theta_{21} = \theta_{22}$, then Eq. (13.61) has no solution.

13.7 Position and Velocity Vectors in Polar Coordinate Frame

As the platform frame is inertial system and its origin is located at the center of the Moon, then we can transform position and velocity vectors' components from the polar coordinate frame ($Or\theta$) to the platform frame, which includes the Moon's center, nominal landing site and the current position and velocity vectors (see Fig. 4.2 of Chap. 4):

$$\begin{aligned}
x &= r \sin\theta, \\
y &= 0, \\
z &= r \cos\theta, \\
v_x &= v_1 \sin\theta + v_2 \cos\theta, \\
v_y &= 0, \\
v_z &= v_1 \cos\theta - v_2 \sin\theta.
\end{aligned} \tag{13.64}$$

where

$$\theta = \tan^{-1}\left(\frac{x}{z}\right).$$

13.8 Platform Frame for a Nominal Landing Site

Consider the Moon-centered Moon-fixed system as a local system with subscript "L". Assume that the longitude and latitude of the nominal landing site in the local system are given as

$$\theta = \theta_{LS}, \qquad \varphi = \varphi_{LS}.$$

Corresponding position vector of the landing site, $\mathbf{R}_{LP}$ in a Moon- fixed system is determined by formulas:

$$\begin{aligned} R_{XLSL} &= R_M \cos\varphi_{LS} \cos\theta_{LS}, \\ R_{YLSL} &= R_M \cos\varphi_{LS} \sin\theta_{LS}, \\ R_{ZLSL} &= R_M \sin\varphi_{LS}, \end{aligned}$$

where R_M is the magnitude of position vector of the landing site in the Moon-fixed coordinate system with the origin at the center of the Moon. R_M is computed based on the assumption with respect to one of the representations of the lunar topographic surface, such as spherical, oblate ellipsoidal or harmonic. The position vector of the landing site, $\mathbf{R}_{LSP}$ can be computed using the formulae:

$$\mathbf{R}_{LSP} = \mathbf{T}_L^P \mathbf{R}_{LSL},$$

where

$$\mathbf{T}_L^P = \mathbf{T}_I^P \left(\mathbf{T}_I^L\right)^T.$$

The transformation matrices $\mathbf{T}_I^P$ and $\mathbf{T}_I^L$ have been defined in Part 1 of the monograph.

13.9 Position and Velocity Vectors in Platform Frame

These vectors in platform frame are expressed using the transformation matrix from guidance to platform frames including the landing site coordinates and counting for the rotation of the Moon (see Fig. 4.2 of Chap. 4):

$$\mathbf{R}_{IP} = \mathbf{T}_G^P \mathbf{R}_{IG} + \mathbf{R}_{LSP},$$

$$\mathbf{V}_{IP} = \mathbf{T}_G^P \mathbf{V}_{IG} + \omega_P \times \mathbf{R}_{IP}. \tag{13.65}$$

where $T_P^G = [a_{ij}]$ is the transformation matrix, $\mathbf{R}_{LSP}$ is the vector of landing site in platform frame. This completes the construction of the BP nominal extremal trajectory from the entry orbit to the BP-final point (the AP-initial point). These procedures show that no repeated AP- and BP- targeting and state vector iterations are needed to construct the BP extremal trajectory. This is the significant advantage of the proposed solutions. As mentioned in the previous subsection, only one iterative process to solve the transcendental equation (see Eq. (13.45)) to determine the thrust percentage is needed to implement the solutions. Of course, this equation may be solved again when the re-targeting takes place. Equations (13.41), (13.45) and (13.48) provide the continuity of the position and velocity vectors at the junction point which connects the end point of the BP-CT arc and the BP-quartic. The analytical solutions in Eq. (13.35) and satisfaction of the boundary and continuity conditions at the TCR point demonstrate the utility of the proposed BP-CT arc in the design of a powered descent trajectory as an advanced alternative to the Apollo-era trajectory and guidance design.

References

1. Klumpp AR (1971) Apollo lunar-descent guidance. Apollo guidance navigation control, R-695. Massachusetts Institute of Technology, MIT Instrumentation Laboratory Report, Cambridge, MA
2. Project Apollo Coordinate System Standards (1965) Apollo program, SE 008-001-1, NASA office of manned space flight. http://ntrs.nasa.gov/archive/nasa/casi.ntrs.nasa.gov/19700076120_1970076120.pdf
3. Lawden DF (1963) Optimal trajectories for space navigation. Butterworths, London
4. Vallado DA (1997) Fundamentals of astrodynamics and applications. McGraw-Hill Companies Inc.

Chapter 14
Simulations of Braking Phase Using Analytical Solutions

14.1 Description of Simulations

The simulations aimed to analyze the analytical solutions for the constant thrust (CT, maximum thrust (MT) if $p_1 = 1$) arc that would transfer the lander from the initial orbit to the throttle recovery point (TRP) followed by the throttle recovery (TR) phase which ends at the BP terminal point. The studies included three parts of the trajectory: (1) from entry orbit to TRP; (2) from TRP to AP initial point (BP quartic); (3) from AP initial point to the terminal descent (TD) initial point (AP quartic).

The following set of input parameters with typical values not representing any specific mission have been selected:
Set 1:

- entry orbit: $e = 0.0290778$, $p = 1803973.415$ m.. The orientation of the orbit is determined by the solution process: $\omega = 1.97294906$ rad; (Orbit 1);
- engine parameters: $c = 3150$ m/s, $\beta_{max} = 16.6$ kg/s.
- initial mass and mass estimate at the terminal point of the braking phase: $M_0 = 14800.000$ kg; $M_{BRF} = 8670$ kg.
- altitude of the initial point of the braking phase and altitude of apogee of the entry orbit: $h_0 = 15000.000$ m., $h_a = 120000.000$ m.

The studies have show that the thrusting trajectory from the initial orbit to the initial point of throttle recovery phase can be described once the integration constants are determined.

Although for the illustration purposes the simulations have been conducted for the case when $p_1 = 1$, the utility of the case with $p_1 < 1$ is of great interest due to the fact that $93\% - 94\%$ of maximum allowable thrust was considered as the highest percentage of thrust used on the braking phase. The braking phase described by the MT arc solutions is shown in Fig. 14.1. The last two parts of the trajectory can be represented by a quartic formulation. The simulation results describing all

© The Author(s), under exclusive license to Springer Nature Switzerland AG 2026

D. M. Azimov and R. H. Bishop, *Apollo-Class Targeting and Guidance for Lunar Powered Descent and Precision Landing*, Springer Aerospace Technology,
https://doi.org/10.1007/978-3-031-91088-3_14

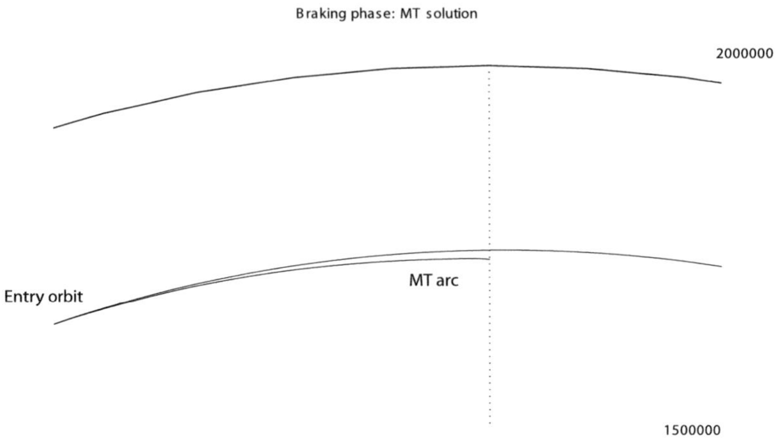

Fig. 14.1 The braking phase replaced by MT arc

three parts of the descent trajectory, including the altitude, the components of the position, velocity and acceleration vectors for each part of the trajectory are given in Figs. 14.2, 14.3, 14.4, 14.5, 14.6 and 14.7.

The studies have shown that considering the trajectory in these three parts mentioned above considerably simplifies the targeting procedures for both quartic phases; It should be noted that the entry orbit, initial altitude, initial mass and mass flow-rate, altitude and velocity at the TR point, exhaust velocity and percentage of thrust should satisfy the transcendental equation, Eq. (13.45). Total acceleration and thrust acceleration vs. time and the thrust angle's behavior on CT (MT) arc are given in Figs. 14.8 and 14.9. All values for the position, velocity, mass, mass-flow rate, exhaust velocity and time are typical or correspond to the Apollo guidance performances (see Refs. [1–3]). Note that the targets for the quartic trajectories are the terminal (final) points of the corresponding phases. If it is required to construct a BP- or AP- quartic, then it should be noted that not all typical values of parameters mentioned in Klumpp's will work to satisfy the given constraints. For example, if the BP-quartic is constructed backwards from its terminal point to the TR point, then the magnitude of the position vector should be continuously increased and the radial velocity should have a positive sign all the time. But this is possible only if the corresponding components of acceleration, jerk and snap vectors have corresponding signs. This type of requirement imposes additional expressions to be satisfied by the quartic parameters. Consequently, the permissible regions for the parameters can be determined.

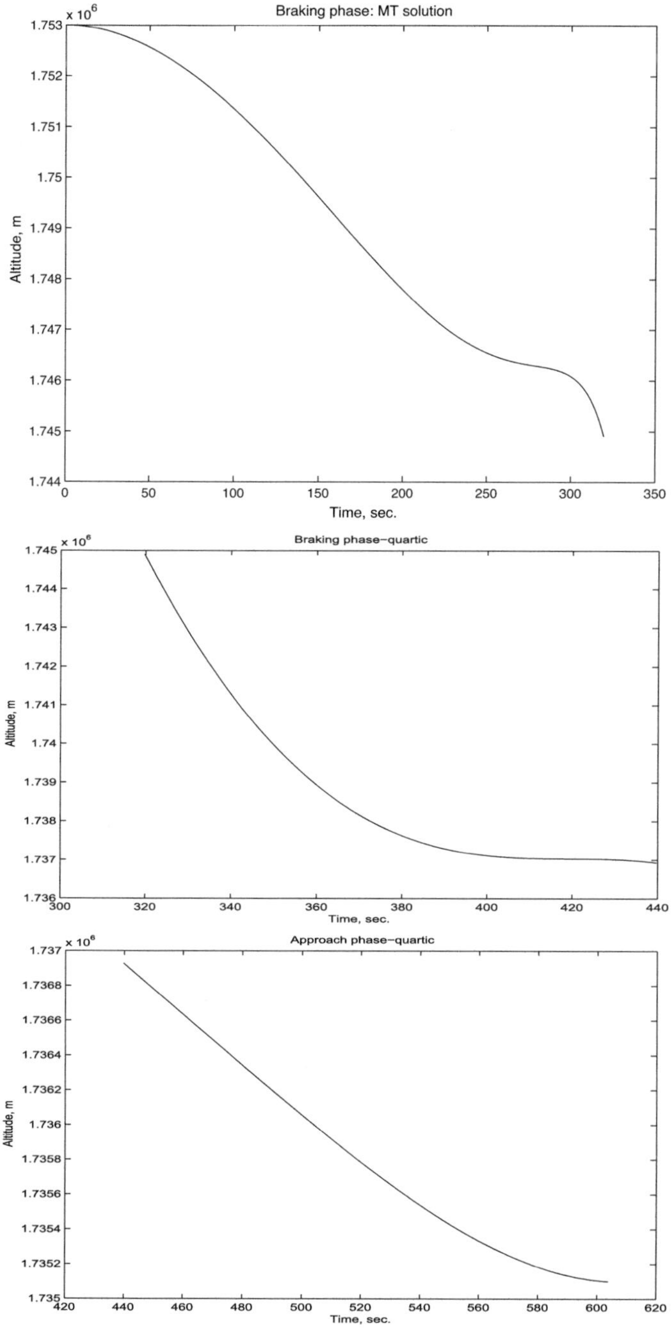

Fig. 14.2 Altitude versus time

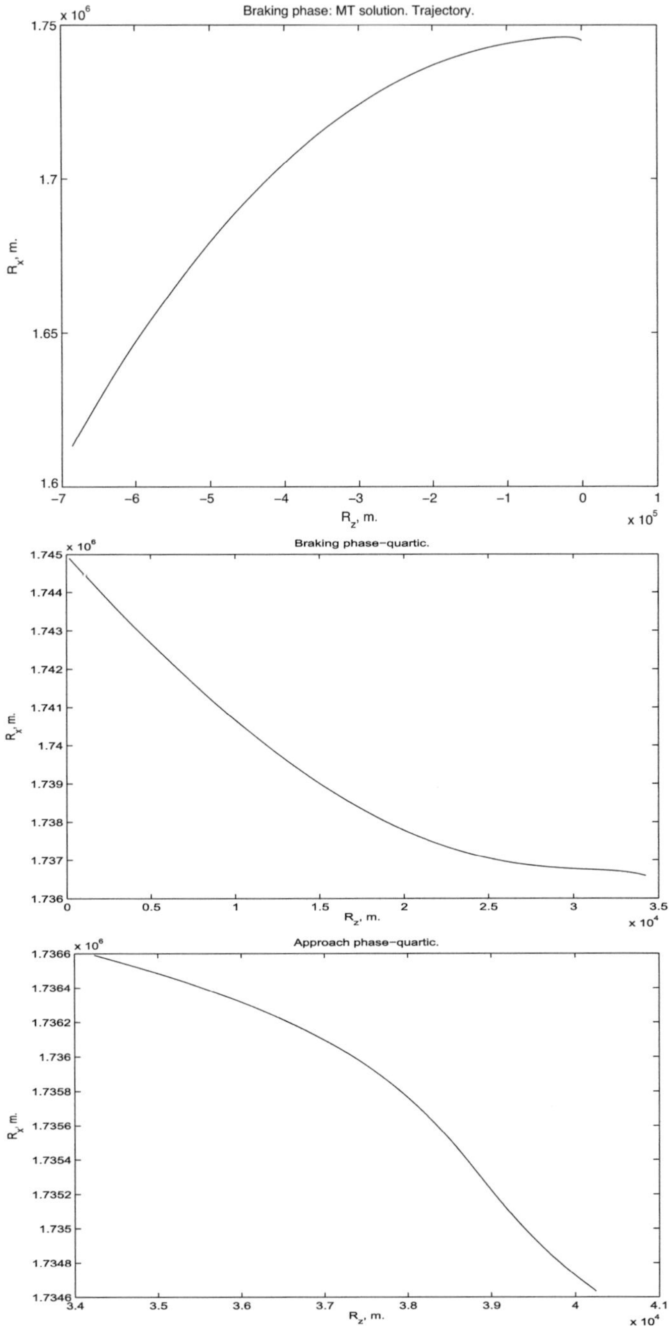

Fig. 14.3 Components of the position vector

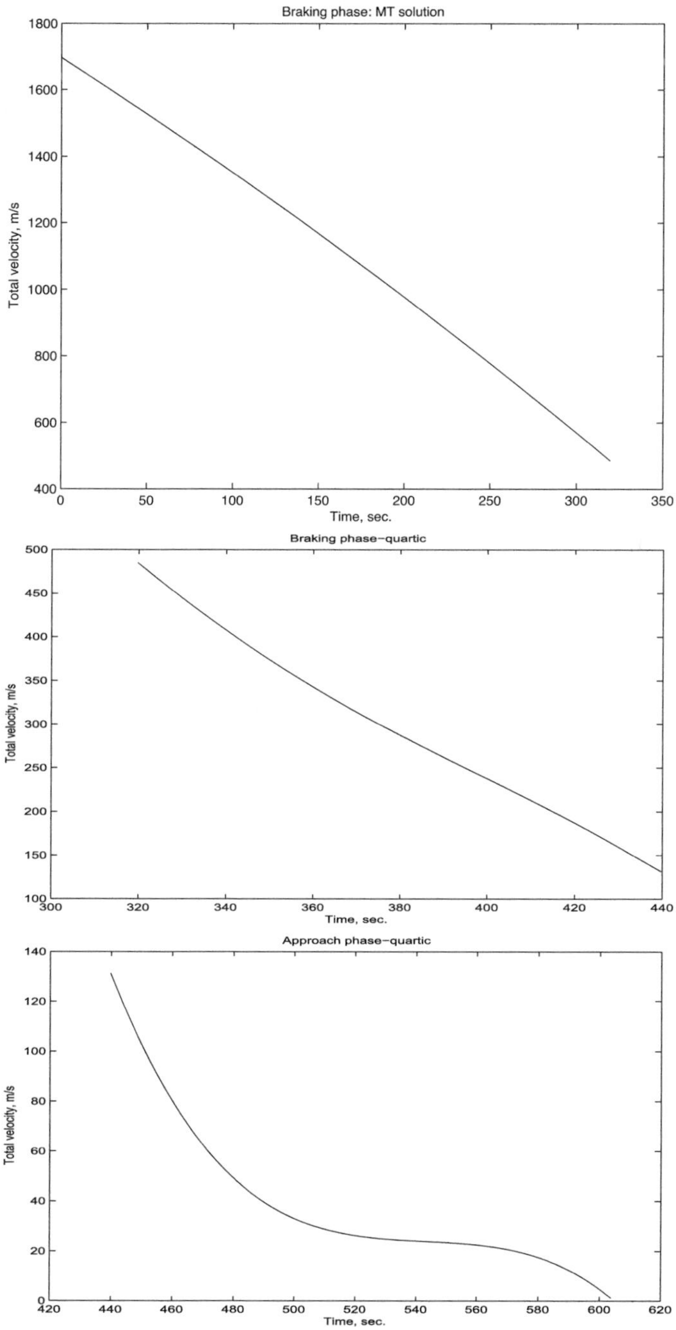

Fig. 14.4 Magnitude of velocity vector

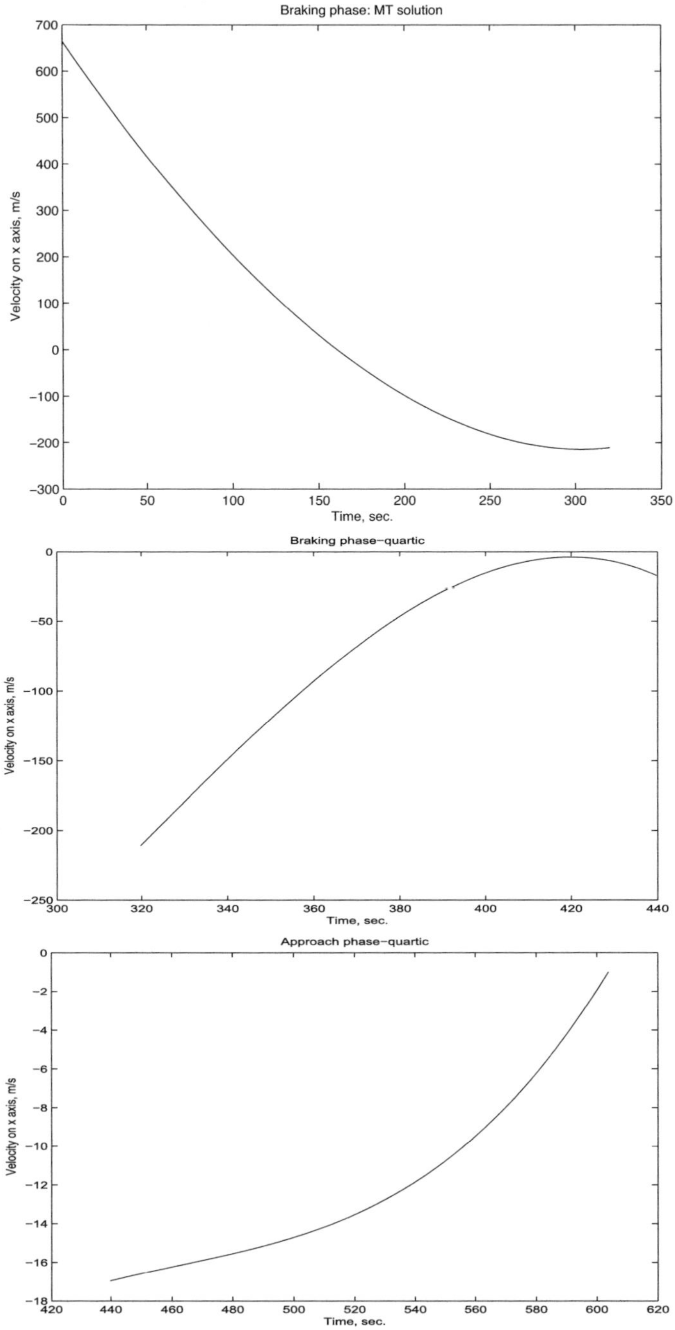

Fig. 14.5 Vertical component of velocity vector, x-axis

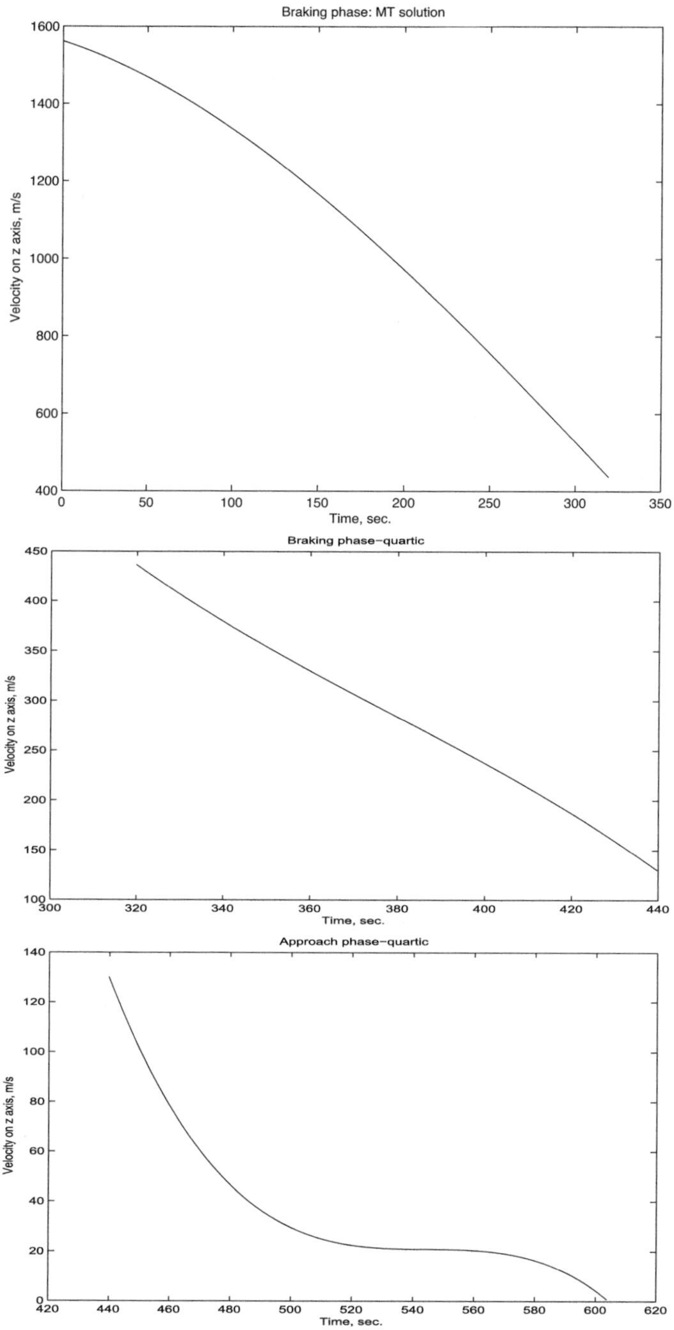

Fig. 14.6 Vertical component of velocity vector, z-axis

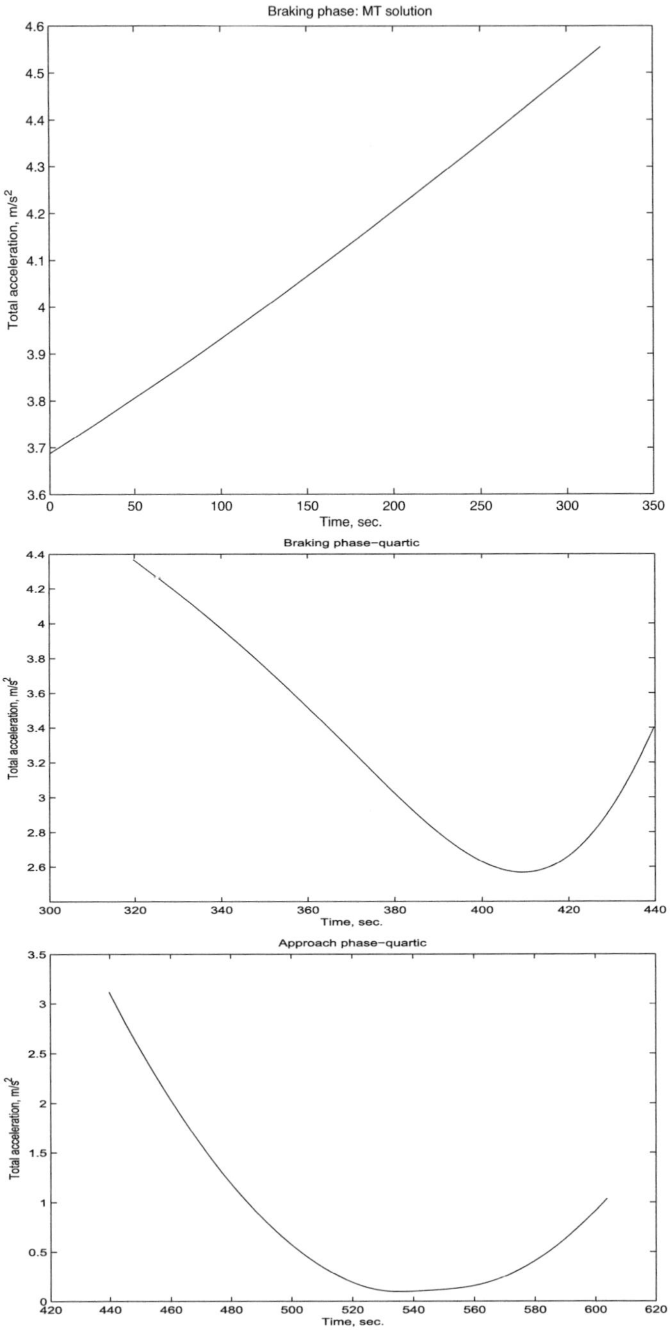

Fig. 14.7 Total acceleration versus time

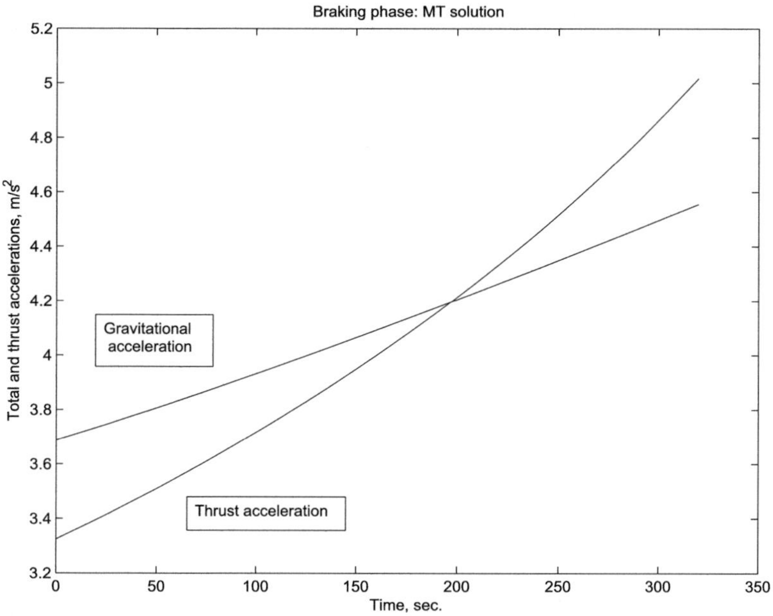

Fig. 14.8 Total acceleration and thrust acceleration versus time on MT arc

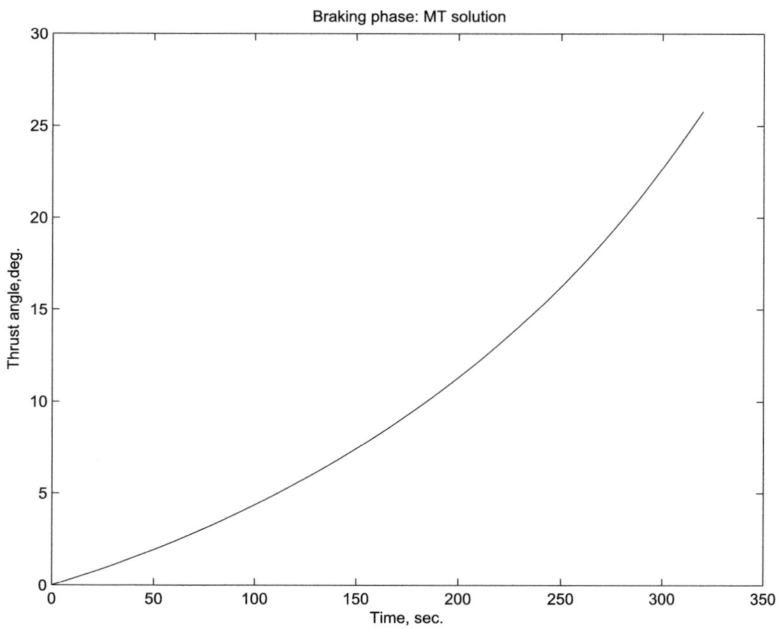

Fig. 14.9 Thrust angle versus time

14.2 Profiles of Thrust, Velocity and Flight Duration

14.2.1 Simulation Inputs

Another two sets of simulations have been conducted to determine the profiles of thrust percentage, velocity of motion and duration of flight on MT arc utilizing the analytical solutions for these arcs. In the simulations, the trajectories started from different altitudes using different entry orbits. It should be noted that the entry orbit, initial altitude, initial mass and mass flow-rate, altitude and velocity at the throttle down point, exhaust velocity and percentage of thrust should satisfy the transcendental equation Eq. (13.45).

The following two input sets of parameters have been selected:
Set 1:

- entry orbit: $e = 0.0290778,\ p = 1803973.415$ m.. The orientation of the orbit is determined by the solution process: $\omega = 1.97294906$ rad; (Orbit 1);
- engine parameters: $c = 3150$ m/s, $\beta_{max} = 16.6$ kg/s.
- initial mass and mass estimate at the terminal point of the braking phase: $M_0 -$ 14800.000 kg; $M_{BRF} = 8670$ kg.
- altitude of the initial point of the braking phase and altitude of apogee of the entry orbit: $h_0 = 15000.000$ m., $h_a = 120000.000$ m.

The values shown above are typical and do not represent any specific mission.

Set 2:
$e = 0.0266312,\ p = 1799875.788$ m, $\omega = 1.97295$ rad (Orbit 2);
$h_0 = 15186.4$ m, $h_a = 111120$ m, $R_{TDP} = 1743213.074$ m;
$M_0 = 15121$ kg, $M_{BRF} = 8662.594$ kg;
$c = 3150$ m/s, $\beta_{max} = 16.6$ kg/s.

The second set of the data mostly corresponds to Apollo 12 characteristics and partially used by Yang [2]. The results show that for a given altitude R_{TDP} at the throttle down point, there exists a certain set of values of V_{TDP} with corresponding duration of MT and the thrust percentage. For the altitudes selected, the thrust percentage can vary from 63% to 154%. Some relationships in this regard are illustrated in Figs. 14.10, 14.11, 14.12, 14.13, 14.14 and 14.15. Analysis show the necessity of additional studies in order to determine the regions of permissible values for the trajectory and guidance parameters mentioned above.

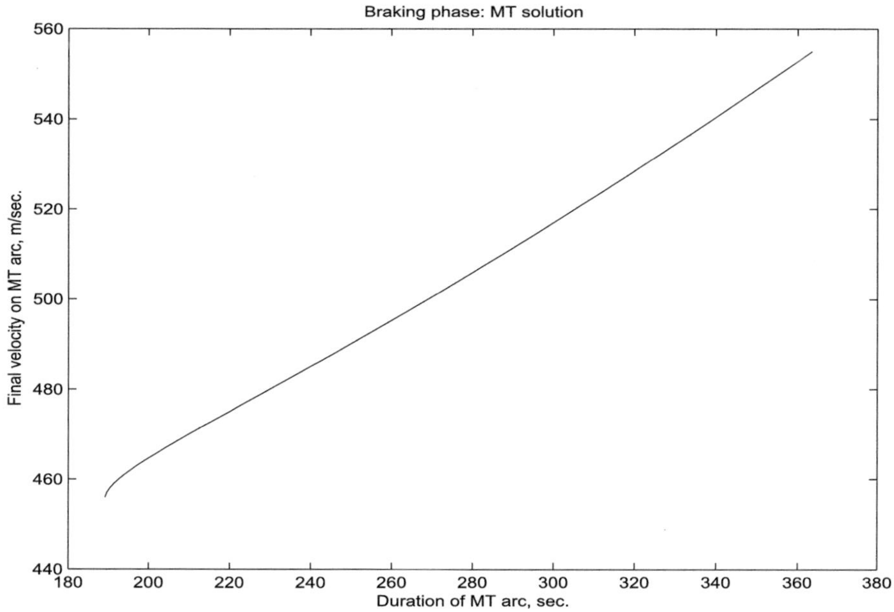

Fig. 14.10 Final velocity versus duration of MT arc. Orbit 1

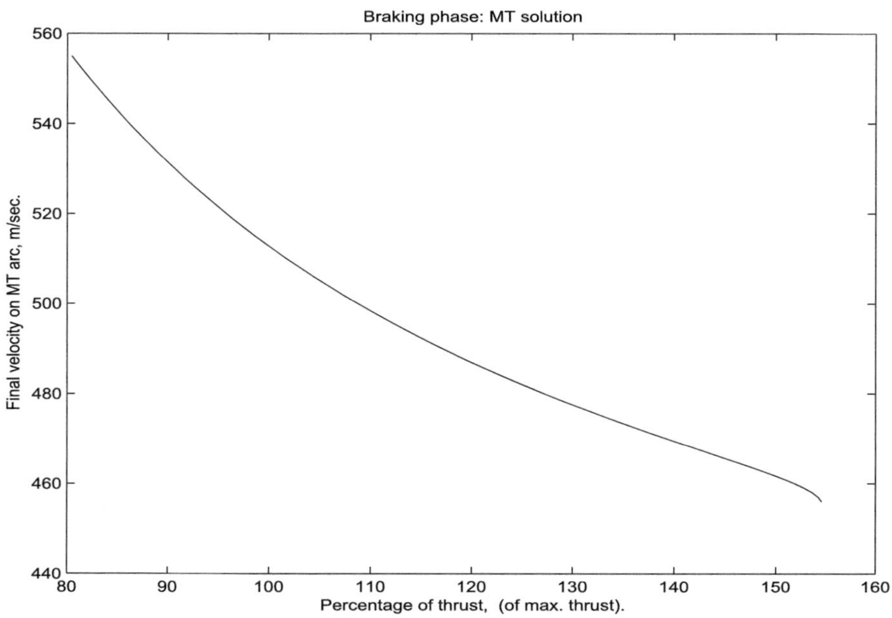

Fig. 14.11 Percentage of thrust versus Final velocity on MT arc. Orbit 1

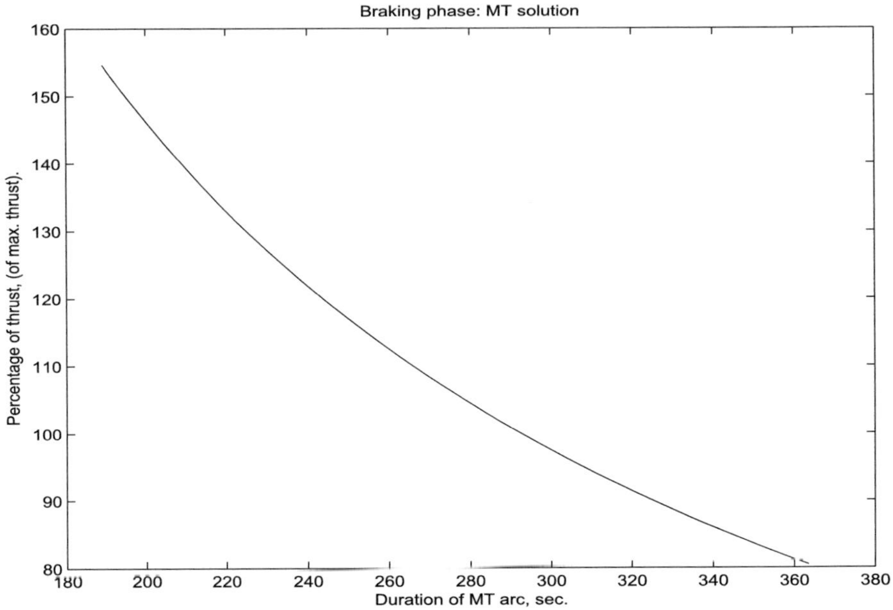

Fig. 14.12 Percentage of thrust versus duration of MT arc. Orbit 1

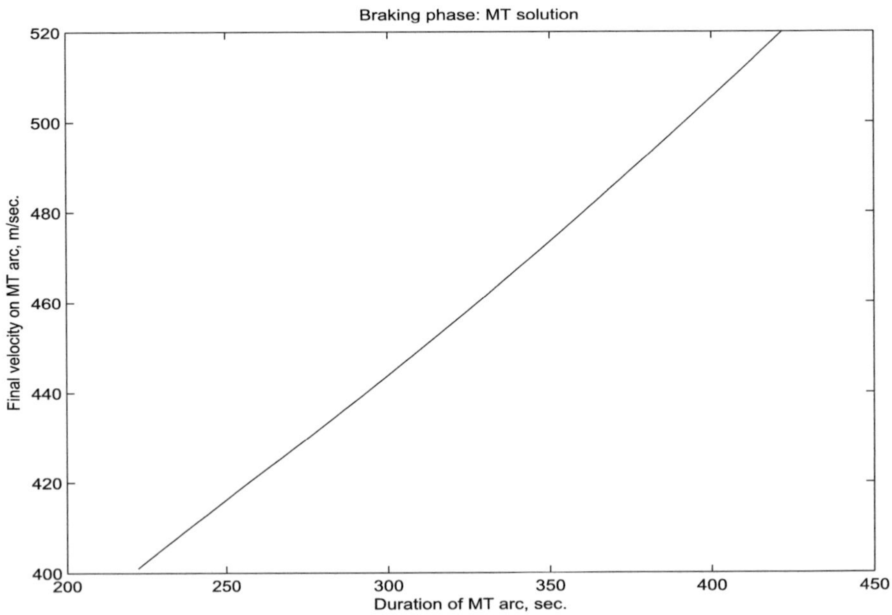

Fig. 14.13 Final velocity versus duration of MT arc. Orbit 2

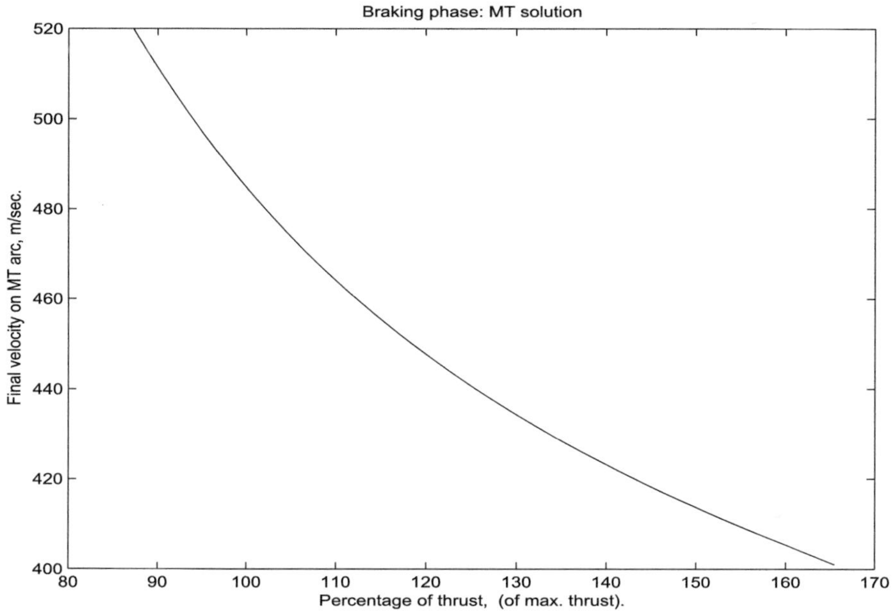

Fig. 14.14 Percentage of thrust versus Final velocity on MT arc. Orbit 2

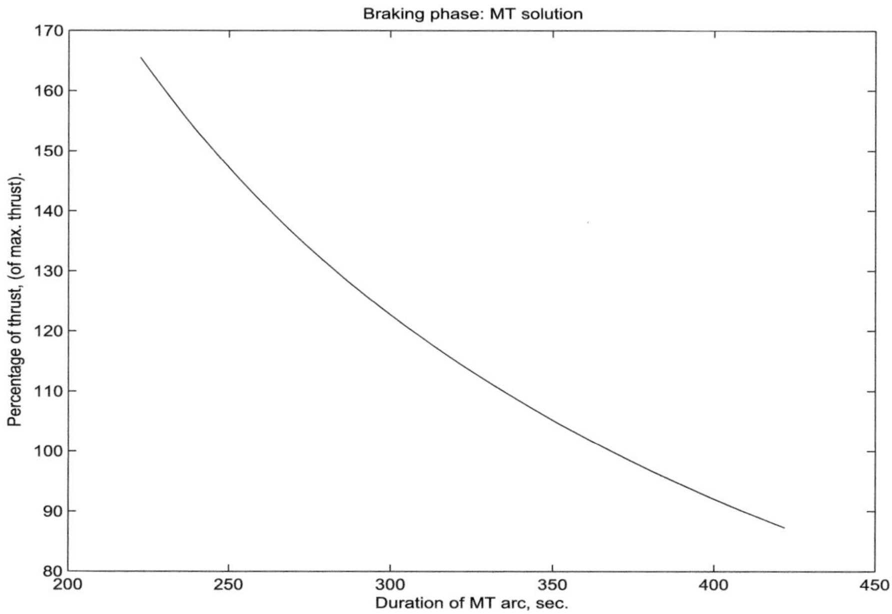

Fig. 14.15 Percentage of thrust versus duration of MT arc. Orbit 2

14.3 Sensitivity of Descent Trajectory to Initial Conditions

In addition to the simulations described in the previous sections, one more set of simulations has been conducted to analyze the sensitivity of the shape of the descent trajectory to the initial conditions. This set of simulations also included three parts of the trajectory: (1) from entry orbit to TRP; (2) from TRP to AP initial point (BP-quartic); (3) from AP initial point to the terminal descent (TD) initial point (AP-quartic). The inputs of the simulations are as follows:

Set 3: $e = 0.0268649,\ \ p = 1800494.606$ m, $\omega = 1.65$ rad (Orbit 3);
$h_0 = 15390$ m, $h_a = 112200$ m, $R_{TDP} = 1744900$ m;
$M_0 = 14300$ kg, $M_{BRF} = 8670$ kg;
$c = 3150$ m/s, $\beta_{max} = 15.1$ kg/s;

The simulations show that the shape of MT arc is highly dependent on initial mass, exhaust velocity and mass-flow rate, and not highly sensitive to changes in perilune and apolune of the entry orbit. Unlike the MT arc, the BR and AP quartics have not changed their shape, including monotonic character of the altitude on these quartics (see Figs. 14.16, 14.17, 14.18, 14.19, 14.20 and 14.21). The qualitative behavior of the velocity and acceleration vectors, and the thrust percentage with respect to duration of the MT arc and final velocity at TDP remained the same compared to the analysis with previous sets of initial conditions.

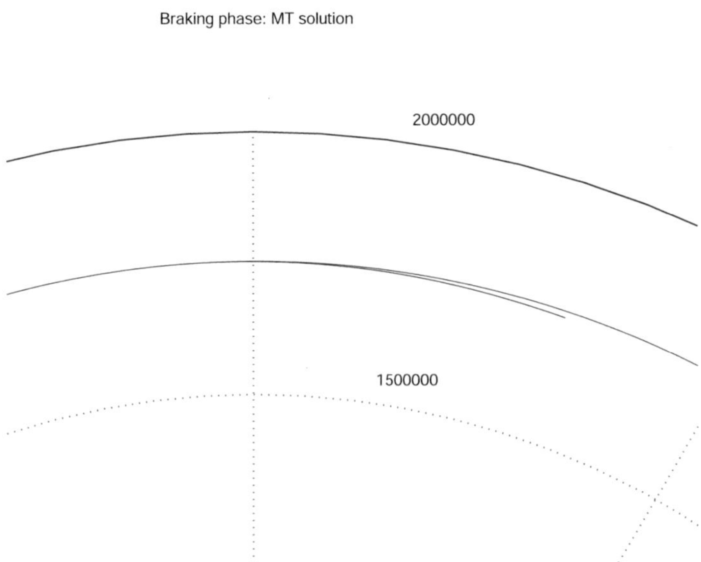

Fig. 14.16 The braking phase replaced by MT arc. Set 3

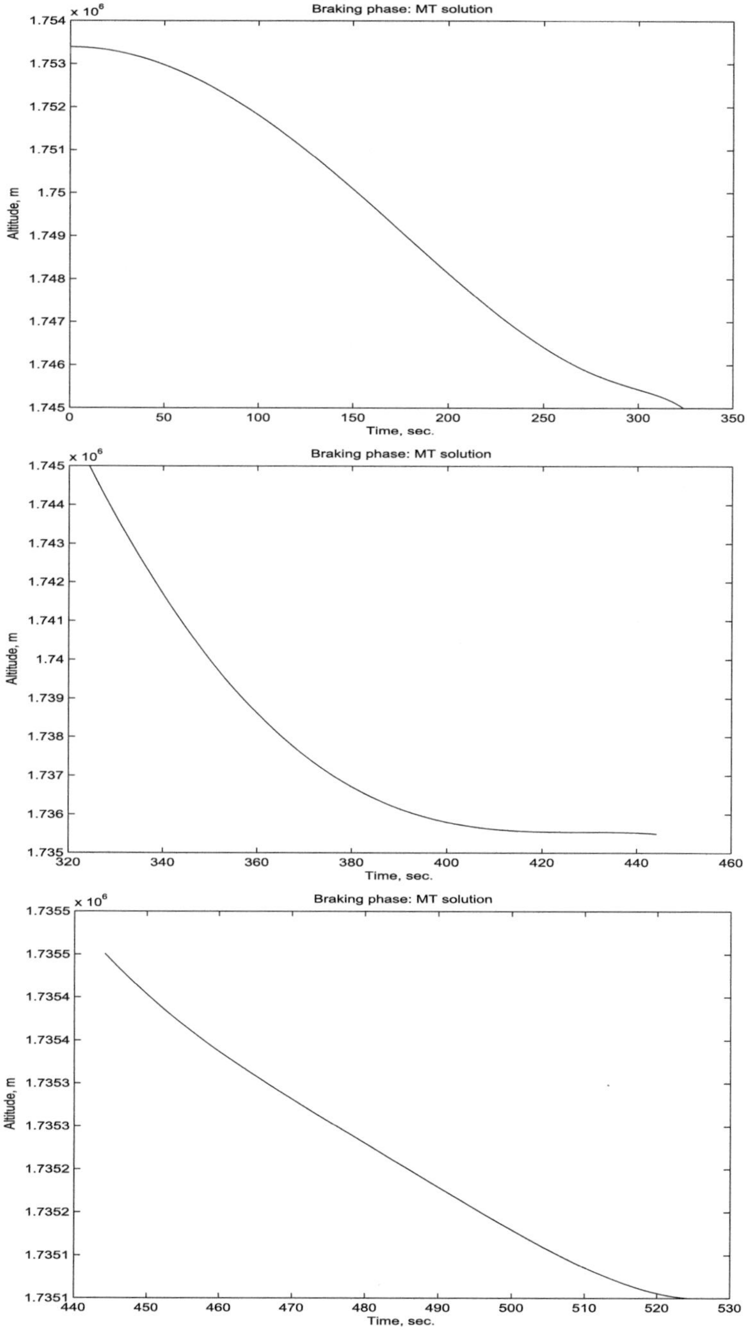

Fig. 14.17 Altitude versus time. Set 3

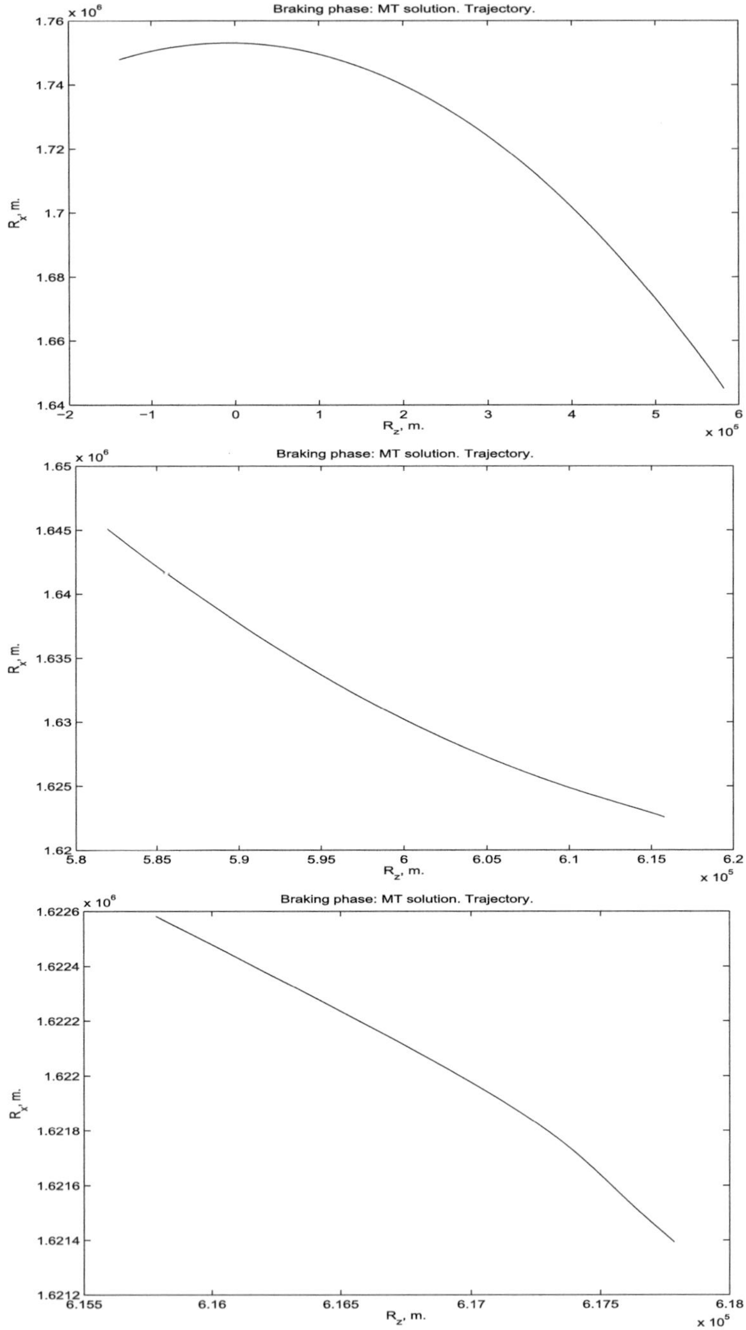

Fig. 14.18 Components of the position vector. Set 3

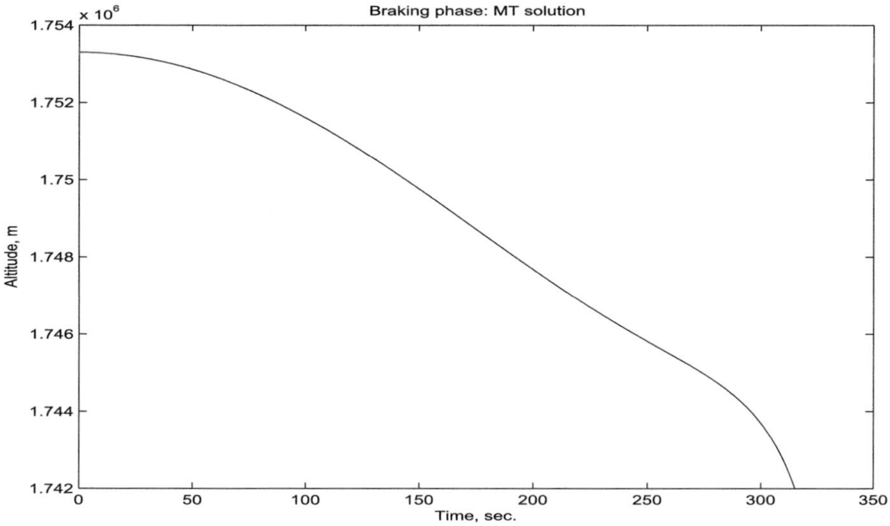

Fig. 14.19 Sensitivity of altitude on MT arc to initial mass: $R_{TDP} = 1742000$ m, $h_0 =$ 15300 m., $M_0 = 14850$ kg. Set 3

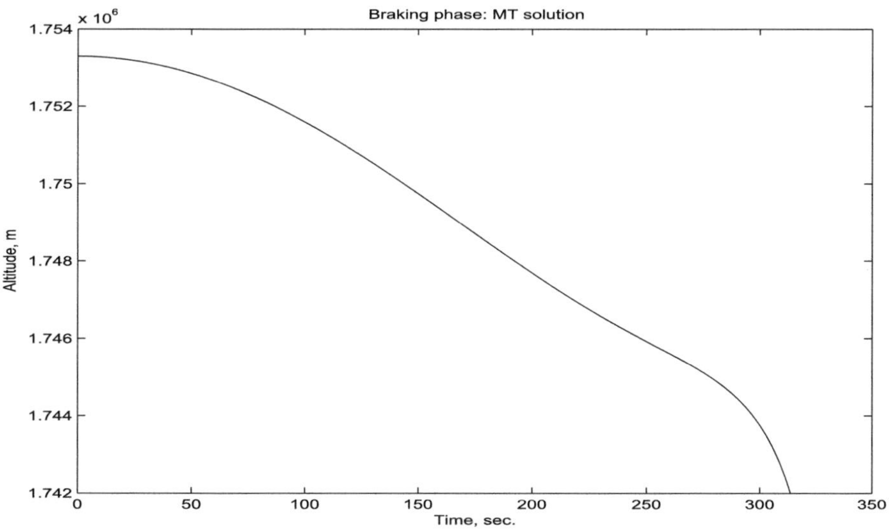

Fig. 14.20 Sensitivity of altitude on MT arc to initial mass: $R_{TDP} = 1742000$ m, $h_0 =$ 15300 m., $M_0 = 14700$ kg. Set 3

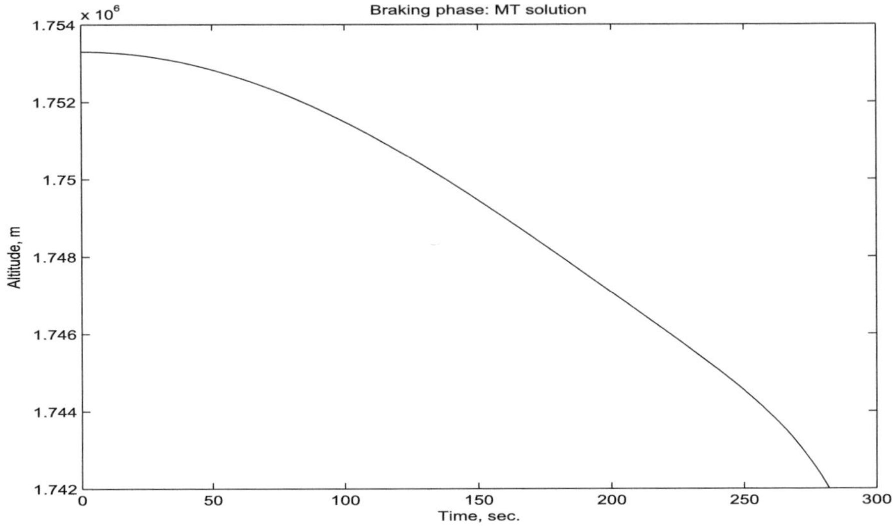

Fig. 14.21 Sensitivity of altitude on MT arc to initial mass: $R_{TDP} = 1742000$ m, $h_0 = 15300$ m., $M_0 = 14200$ kg. Set 3

14.4 Simulations of Powered Descent at Moon South Pole

In this chapter, a powered descent maneuver at the Moon South Pole is simulated using the analytical solutions for the CT (MT) arcs. The inputs to BP- and AP- targeting programs, the DPS parameters, and the entry orbit are considered similar to those used in the simulations of the Apollo missions. The landing site (LS) coordinates were selected based on the known location of the Shackleton Crater, which lies at the South Pole [4]. The epoch 1.0 h 0.0 m 0.0 s of January 1, 2016 has been considered as the nominal landing time. The simulation results reflect the similar qualities and profiles of the parameters of the Apollo guidance performances. Design of targeting and guidance for the South Pole mission can be developed similar to the design of Apollo guidance programs. The simulation results are provided in Figs. 14.22, 14.23, 14.24 and 14.25.

14.5 Concluding Remarks

The Apollo BP- targeting program has been explained and reconstructed. The main part of this program has been modified by applying the closed-form solutions. Application of these solutions enables implementation of targeting algorithm on-board as many times as needed without engaging iterative procedures. This allows for the algorithm to enable the performance of various autonomous operations, such as mapping

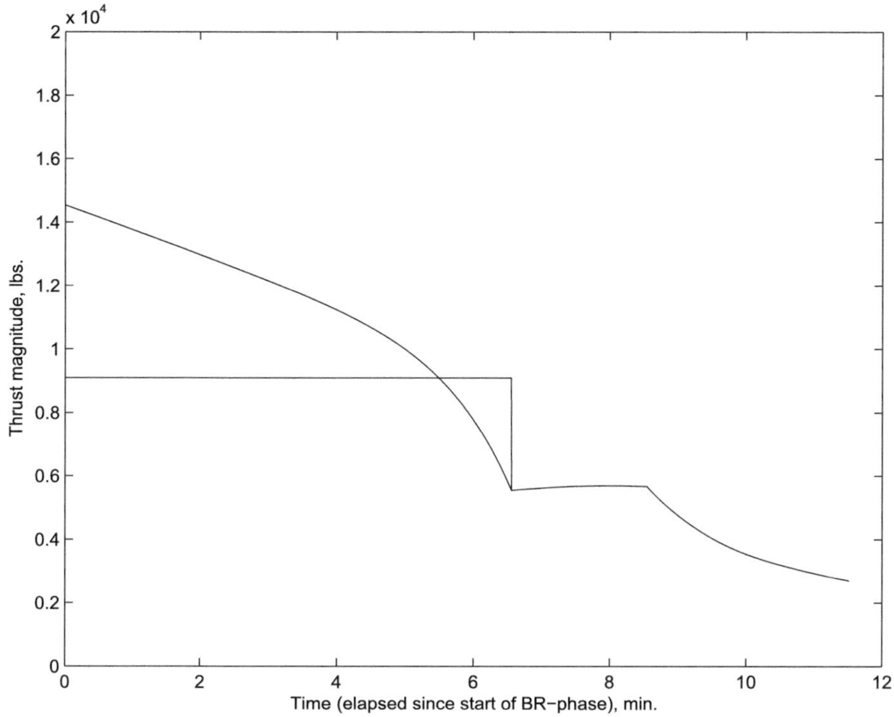

Fig. 14.22 Actual and commanded thrust profiles

of the terrain features, hazard detection and avoidance, and re-designation of LS. The terminal conditions and the mission-specific constraints can be satisfied by utilizing the arbitrary integration constants thereby providing the robustness of the algorithms to all mission-specific requirements.

The part of the braking phase starting from the initial orbit and ending with the initial point of the throttle recovery interval (throttle down point) has been modified by the closed form extremal CT (MT) arc solution. The guidance scheme for this part can be modified based on these solutions. It has been shown that this modification avoids the iterations and convergence issues, and successfully solves the BP and AP targeting problems.

The programs have been validated and evaluated by comparing the results of the simulations to the Apollo 11 and 12 landing missions. As shown in Part 1 of the monograph, the results of the targeting and guidance programs, especially the thrust, pitch angle, velocity and range profiles are very similar to those of these Apollo missions.

From the third set of simulations one can conclude that the shape of MT arc is highly dependent on the values of initial mass, exhaust velocity and mass-flow rate, but the shapes of the BP- and AP- quartics are not sensitive to minor changes in the initial conditions.

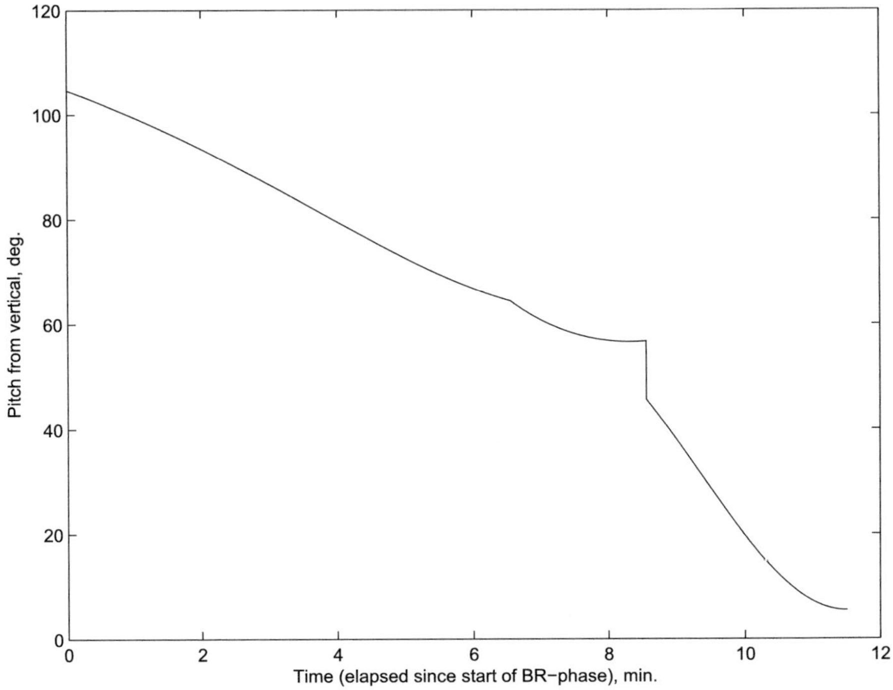

Fig. 14.23 Pitch from vertical

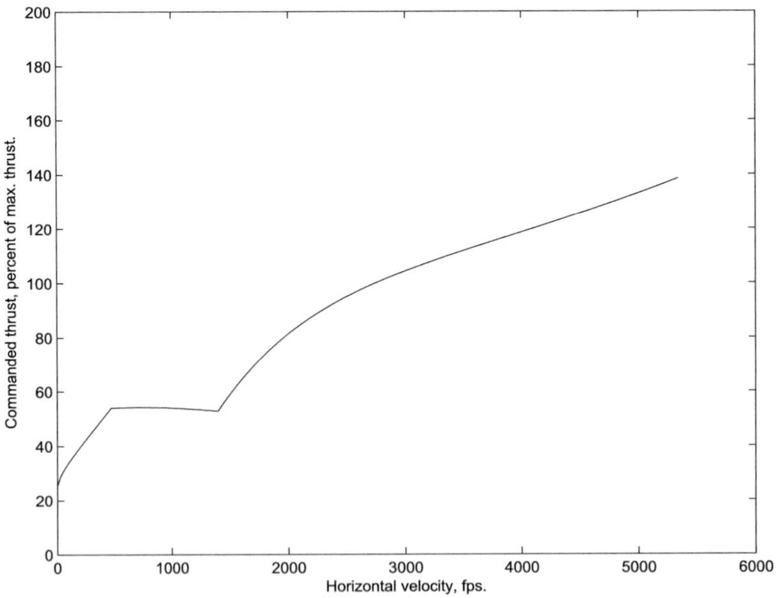

Fig. 14.24 Commanded thrust versus horizontal velocity

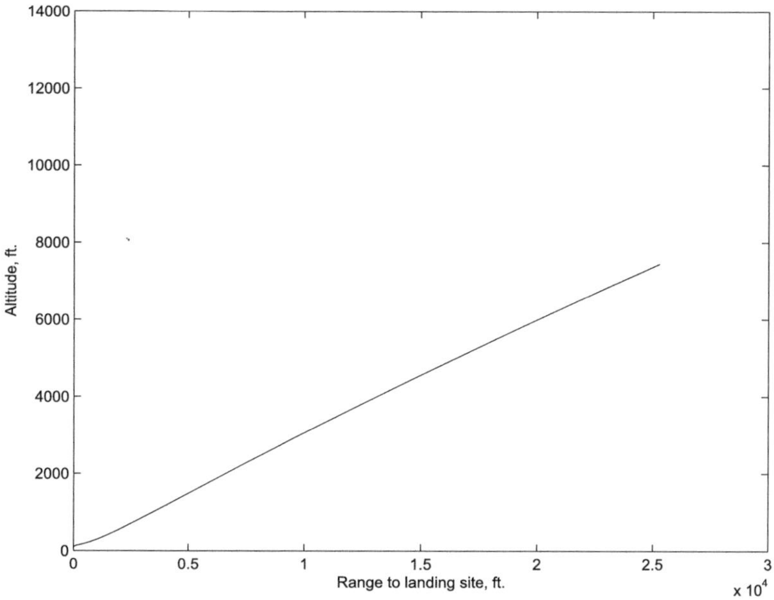

Fig. 14.25 Approach phase: altitude versus range

The basic conclusion of the study is that the Apollo targeting and guidance problems can be solved entirely in explicit and closed form. The solutions presented above do not involve uncertainties or iterations except for the transcendental equation (see Eq. (13.45)), which requires only a few iterations. Also demonstrated in the next chapter are the new targeting results for the South Pole (Shackleton crater) mission. The advanced targeting algorithms described in the previous chapters have demonstrated the feasibility and technical likelihood of success of the future work.

The following efforts would of great importance in further extending and enhancing the Apollo-class guidance and targeting program:

- Exploring the quartic properties. The studies may focus on the development of operational procedures and properties of the quartic formulations according to mission-specific requirements.
- Re-consideration and modification of the Apollo attitude maneuver and throttle routines.
- Development high fidelity simulation model.

References

1. Klumpp AR (1971) Apollo lunar-descent guidance. In: Apollo guidance navigation control, R-695. Massachusetts Institute of Technology, MIT Instrumentation Laboratory Report, Cambridge, MA
2. Yang TL (1971) A targeting scheme for fuel optimal rocket trajectories with applications to the LM descent braking phase. Technical memorandum TM—71-2014-1, Bellcomm
3. Klumpp AR (1968) Manually retargeted automatic landing system for the lunar module (LM). AIAA J Spacecr Rocket 5(2):129–138
4. https://en.wikipedia.org.wiki.Shackleton (crater)

Chapter 15
Real-Time Targeting and Re-Targeting

15.1 Concept of Real-Time Targeting and Re-Targeting

Real-time targeting implies on-board computations of the target states, the LS coordinates and the re-orientation of the nominal $X_p - Z_p$ plane. Integrated real- time targeting and re-targeting procedures are an integral part of an on-board targeting and guidance subsystem. Advantages of the existence of the closed-form solutions and the possibility of determining the elements of the transformation matrix, $\mathbf{T}_P^L$ from the platform system to the local system in an explicit form allow us to perform real-time targeting at any given time on the descent trajectory. Specifically, the integration constants will change due to the change in the target (or final) parameters, and this will yield a new trajectory starting from the same instant of time. Note that the original design of Apollo landing trajectories included the BP-quartic and AP-quartic, and the utility of these quartics was based on the ground-based targeting (see Fig. 15.1). The general concept of real-time targeting can also be developed based on the utility of a quartic polynomial and it can be observed that the change in the targets will also cause a change in other target parameters for the rest of the trajectory. This may also cause change in the nominal landing time and in matrices $\mathbf{T}_P^L$ and $\mathbf{T}_P^I$. The resulting new nominal $X_p - Z_p$- plane will contain three points: the landers center of mass, new LS, and the Moon's center of gravity. In particular, the change of the LS coordinates will cause changes in all targeting parameters of the rest of the trajectory. The re-orientation of the $X_p - Z_p$- plane is the main characteristic of the real-time targeting procedure as it contains the nominal re-targeted trajectory, the "heart" of the guidance system, described by the proposed BP-CT arc solutions. The concept of real-time targeting and re-targeting is illustrated in Fig. 15.2. Assume that at time t_c the LS vector, $\mathbf{r}_{ls}$ is changed as a result of image data processing, or by recommendations from the ground control system or according to the commands from the targeting and guidance subsystem. This may also cause a change in the nominal landing time, t_f^0. As $\mathbf{T}_P^L = \mathbf{T}_P^L(\mathbf{r}_{lsl}(\omega_l))$, the elements of $\mathbf{T}_P^L$, the new target state, re-designation of LS, re-computation of nominal landing time and re-initialization

© The Author(s), under exclusive license to Springer Nature Switzerland AG 2026

D. M. Azimov and R. H. Bishop, *Apollo-Class Targeting and Guidance for Lunar Powered Descent and Precision Landing*, Springer Aerospace Technology,
https://doi.org/10.1007/978-3-031-91088-3_15

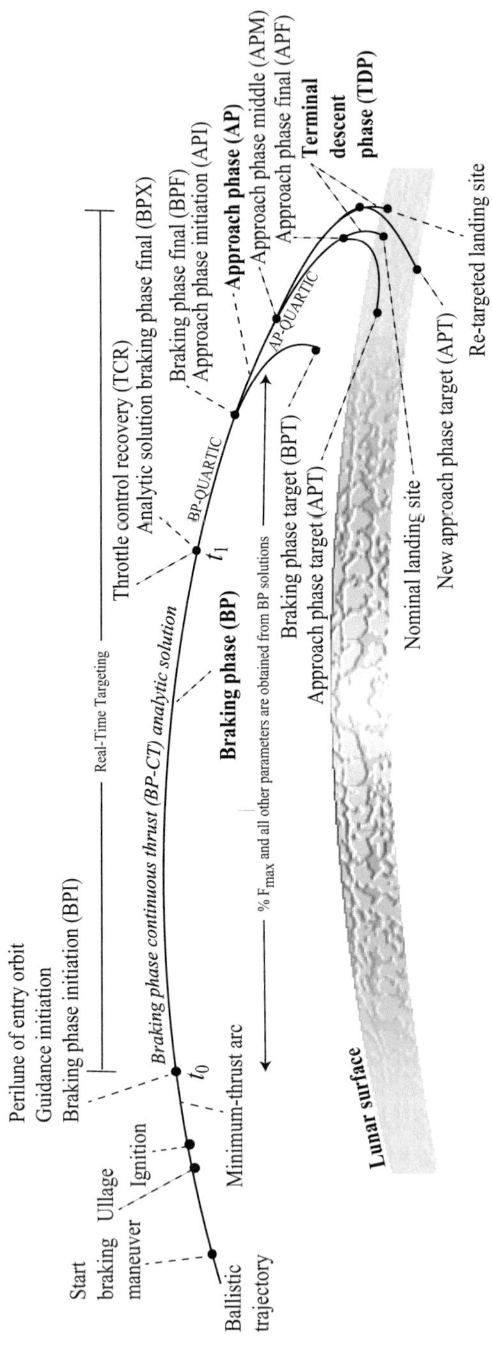

Fig. 15.1 Apollo descent trajectory design with ground-based targeting

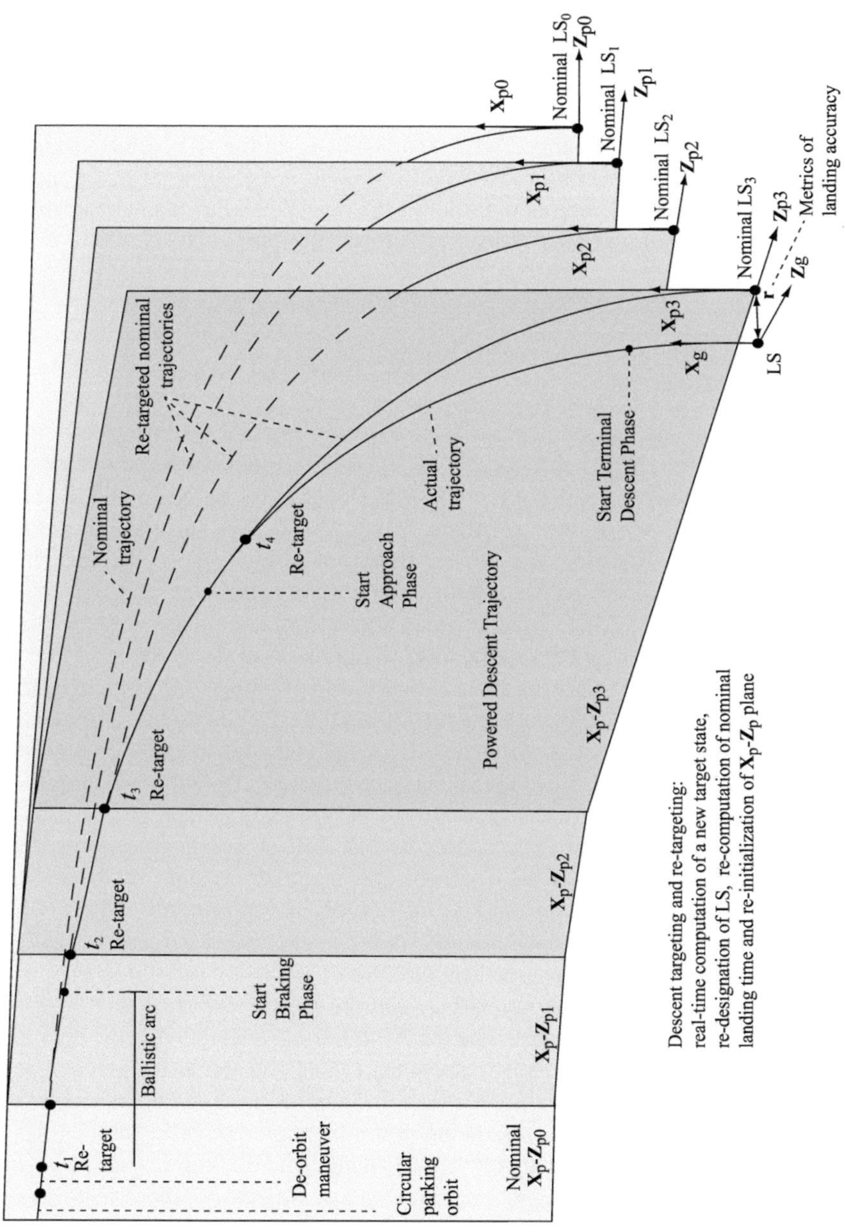

Fig. 15.2 Illustration of the concept of real-time targeting and re-targeting

of $X_p - Z_p$ plane will change accordingly. Here $\mathbf{r}_{lsl}$ and ω_l are the LS-position vector and Moon's rotation vector defined in the local coordinate system. In turn, this causes changes in the elements of $\mathbf{T}^l_p$. Consequently, the $X_p - Z_p$- plane, which contains the nominal trajectory, will change its location, thereby automatically changing the LS location. The new location of this plane will cause changes in all targeting parameters of the rest of the trajectory. Similarly, the re-location of this plane is the result of the re-targeting procedure. The targeting procedure is easier and requires less computations if $\mathbf{r}_{ls}$ is changed when the lander is on the AP. In this case, it is not necessary to solve the continuity equations valid at the BP-final point and the AP-initial point. It will be enough to initialize the new $X_p - Z_p$- plane by changing the target states according to the quartic trajectory solutions.

15.2 Computation of Targeting Parameters

The computation of the targeting parameters depends on the trajectory phase, and it is straightforward as it is based on the BP-CT arc and the quartic solutions. Targeting during the closed-form BP-CT arc includes computation of the position and velocity vectors, the final time and the constants using new vectors, substitution of these parameters into the re-formulated guidance equations and re-execution of guidance. Targeting during BP-quartic consists of the computation of the state vector (position, velocity, acceleration, jerk and snap vectors), thrust angle and mass at the BP-final point, construction of a new BP-quartic, and re-execution of the guidance using the BP- target states. Targeting during the AP-quartic also includes the computation of the position and velocity vectors at the AP-mid and the AP-final points, computation of the new AP-target states, and their substitution in the guidance equation with subsequent execution. The real-time targeting and re-targeting capabilities demonstrate the utility of the proposed closed-form BP-CT solutions in extending the original Apollo ground-based targeting procedures to on-board targeting procedures. Note that the analytical synthesis of the descent trajectory with no state vector and target state iterations can be demonstrated by finding the appropriate values of the integration constants and other parameters (thrust angle, thrust percentage, etc.) to satisfy the boundary conditions (see Eqs. (13.38)–(13.42)), by satisfying the continuity conditions at the junction points between the BP-CT arc and the BP-quartic (see Eq. (13.48)) and the BP-quartic and the AP-quartic. Note also that the only iteration used in the proposed design is to solve Eq. (13.45) for the thrust percentage at given boundary conditions. Only a few iterations may be required to converge to the solution of this equation. In general, the proposed integrated targeting and guidance design is explicit in terms of the parameters used and generic with respect to a planetary body and to the design of a descent and landing trajectory. These features can potentially increase the feasibility of extending the proposed design to future cargo and robotic landing missions.

15.3 Comparison Analysis

For the comparison analysis, the Apollo 12 post-flight trajectory data and a published optimal reference trajectory data can be considered [1]. In Ref. [2], the optimal reference trajectory was designed based on the well known bilinear tangent steering law (uniform gravity field). The values of the position and velocity vector magnitudes, and the mass at the TCR point and at the end point of BP-quartic of Apollo 12, and the optimal reference trajectory were compared to those of the sequence BP-CT arc—BP-quartic (see Fig. 15.1). All given values of the parameters correspond to the Apollo 12 LM's descent trajectory [2]. These parameters are:

$$h_0 = 15186.4 \text{ m}, \; h_a = 111120.0 \text{ m}, \; r_{bi} = 1751100 \text{ m}, \; v_{bi} = 1695.21 \text{ m/s},$$

$$F_{max} = 46704 \; N, \; I_{sp1} = 295.7 \text{ s}, \; I_{sp2} = 304.5 \text{ s}, \; m_0 = 15121 \text{ kg},$$

$$\beta_{max} = 16.6 \text{ kg/s}, \; m_{bf} = 8662.594 \text{ kg},$$

where I_{sp1} and I_{sp2} are the specific impulses at the minimum and maximum thrust levels, respectively. The BP-CT arc data have been computed using the analytical solutions in Eq. (13.35) (see Table 15.1) [1]. Note that $r_b f$ and $v_b f$ for the optimal case were adopted from [2]. The analysis of this data shows that all comparison parameters on the BP-CT arc, except for the velocity magnitude, are close to those of the LM. One of the advantages of the analytical solutions is that for given r and/or v at the TCR point, Eq. (13.45) provides a wide range of p_1 and t_1. In these analyses, for the given $r_{bx} = 1743213$ m and $v_{bx} = 498.0$ m/s, the thrust percentage on the BP-CT arc was found to be $p_1 = 0.947$. Although the BP-CT arc solutions use less than 30 kg more fuel compared to the Apollo 12 LM until the TCR point, the final masses at the BP-final point for these cases are practically the same. Analysis of Eq. (13.45) show that this difference in the fuel consumption at the TCR point is highly dependent on r and v computed at this point, and on m_0. For example, for a given r at this point, the lower values of m_0 lead to the lower values of p_1, v and t_1. As shown in [2], the bilinear tangent steering law and the sine-cosine steering (constant altitude) law performed very close to the optimal reference trajectory: $m_{bf} = 8726.7$ kg for the first law and $m_{bf} = 8726.9$ kg for the second law, respectively. Although for the given set of the

Table 15.1 Apollo 12, optimal ref. trajectory and BP-CT arc data

Points	TCR (BPX)	BPF
Parameters	Apollo 12/Optimal/BP-CT	Optimal/BP-CT
Time since ign., (s)	388.0 345.44 388.67	508 465.44 508.67
Inertial position, (m)	1743230 1733065.43 1743213	173784 173786 1740300
Inertial velocity, (m/s)	426.61 399.18 498.0	132.45 133.0 119.81
Mass, (kg)	9744.09 10601.24 9714.3	8662.55 8726.89 8662.95

parameters, the Apollo 12 LM and the BP-CT arc performances are less fuel-efficient compared to the optimal reference trajectory and to both trajectories designed by using the bilinear tangent and sine-cosine steering laws, the iterative procedures and approximations used in the optimal trajectory solutions may impact the landing accuracy. Note here that due to their explicitness in terms of the design variables and integration constants, the proposed BP-CT arc solutions can provide desired flexibility in improving the landing accuracy by enabling on-board re-targeting. In addition, these solutions provide autonomy of the on-board computations of the target states.

15.4 Simulations of Descent and Landing at Shackleton Crater

As is shown in the previous sections, the advantages of the explicit solutions for the BP-CT arc represent an extension and enhancement of the Apollo ground-based targeting algorithms based on the capabilities of fully autonomous and on-board targeting, re-targeting and automatic re-designation of the LS. Studies have been conducted to analyze the landing maneuver and to design a nominal descent trajectory and to implement the guidance algorithm to bring the lander to a predetermined LS. The sequence of the maneuvers of interest is similar to a typical Apollo landing maneuver and includes the de-orbit maneuver, transfer to the perilune of the entry orbit, descent and landing maneuvers. The design maneuver trajectory consisted of the transfer trajectory from the apolune to the perilune of the elliptical entry orbit, the BP-CT arc, braking, approach and terminal descent phases. Note that the ullage and the trim phases of a typical Apollo descent trajectory are not simulated for simplicity and without loss of generality of the design trajectory. The simulations included computations of the matrices of coordinate transformations and necessary time and mass parameters, computation of the de-orbit maneuver parameters, design of nominal AP-quartic and BP-quartic including their connection to determine the target states, design of the BP-CT arc and the computation of its constants, determination of the TDP guidance parameters, and design of the guidance algorithm. The simulations started with the design of the AP to compute the initial, final and target states of this phase in the guidance coordinates so that the AP constraints are satisfied. The following parameters have to be given to simulate a descent trajectory: nominal landing site (LS) coordinates, nominal landing time, descent engine parameters, the lander's initial mass and the entry orbit parameters. The LS coordinates $(h_l,\ \phi_l,\ \theta_l)$ used in the studies correspond to the Shackleton Crater located at the Moon South Pole. As the lunar cargo and robotic missions are planned for the future, we arbitrarily selected November 14, 2022, 0.0 h 0.0 m 0.0 s as as the nominal landing time. The descent engine parameters are those of a chemical propulsion system that can produce a finite thrust, F with constant specific impulse, I_{sp}. Maximum allowable thrust, F_{max}, and I_{sp} have been selected arbitrarily, but within reasonable bounds.

The entry orbit parameters, h_p and h_a are considered similar to those given in the documents related to the Apollo missions [2, 3].

15.5 Constraints, Inputs and Outputs

The following AP- and BP- constraints are used:

AP-constraints:

$$\begin{aligned} r_{xafg} &= 30.0\ \text{m}; \\ v_{xafg} &= -1.0\ \text{m/s}; \\ r_{xamg} &= 114.0\ \text{m}; \\ v_{xamg} &= -6.5\ \text{m/s}; \\ r_{xaig} &= 2470.0\ \text{m}; \\ T_{ai} &= -165.0\ \text{s}; \\ T_{am} &= -70.0\ \text{s}; \\ T_a f &= -25.0\ \text{s}; \\ \gamma &= 20^\circ. \end{aligned}$$

Here $\gamma = \arctan(x_g/z_g)$. The AP-targeting constraints must be satisfied at the beginning, middle and final points of this phase. Given these parameters, the initial, final and target states of the AP-quartic can be completely determined using the AP-targeting algorithm (see Chap. 5) [4].

BP-constraints:

These constraints are mostly related to the BP-quartic and consist of the final state and the operational constraints. The final state constraints are the following:

$$\begin{aligned} \mathbf{r}_{bfg} &= \mathbf{r}_{aig}, \\ \mathbf{v}_{bfg} &= \mathbf{v}_{aig}, \\ \mathbf{a}_{bfg} &= \left(\frac{F_{bf}}{m_{bf}} \mathbf{u}_{bfg} - \mathbf{g}_m(g_m,\ 0,\ 0) \right), \\ \mathbf{j}_{bfg} &= \left(-(a_{xbfg} + g_m) \frac{\dot{m}_{bf}}{m_{bf}}, 0, -a_{zbfg} \frac{\dot{m}_{bf}}{m_{bf}} \right), \\ \mathbf{s}_{bfg} &= 0 \end{aligned} \tag{15.1}$$

and $\theta_{pbf} = 50^\circ$. The operational constraints include the conditions for the duration of the BP-quartic and the thrust conditions. The BP-quartic is used in the simulations to reduce the initial velocity on the AP, to conveniently determine the thrust vector and provide a fuel economy. These reasons are associated with the fact that the guidance algorithm will determine these parameters in order to satisfy the state and operational constraints. It was assumed that $\Delta t_{cf} = 70.0$ s and $F_{cf} = 0.9F_{max}$. Other constraints are related to the target referenced times: $T_{bi} = -664.4$ s and $TBPF = -0.1$ s.

BP-CT arc constraints:

The BP-CT arc constraints are the initial conditions on this arc: $h_p = 15000.0$ m, $h_a = 100000.0$ m, $\theta_0 = 16°$. The final conditions (the initial conditions of the BP-quartic) are $r_1 = 1741233.18$ m and $v_1 = 362.628$ m/s. No specific constraints are imposed on the TDP.

Inputs. The inputs of the simulations include the trajectory constraints described in this subsection, the initial and final conditions, the mass and engine parameters, the clock times and target referenced times, the LS coordinates and the re-targeting initiation parameters (see Table 15.2). Some of the inputs are typical, do not represent any specific mission and have been adopted from the Apollo- related documents [5]. In Table 15.2, the units of the parameters are as follows: seconds for time, degrees for angles and meters for distance. Note that $\Delta t_{LAND} = \Delta t_f^0$ (see Fig. 15.2). The propulsion system parameters and the mass have the following values: $F_{max} = 100000\ N,\ I_{sp} = 400$ s and $m_0 = 35000$ kg.

Outputs of BP-targeting:

The outputs of BP-targeting algorithm consists of the following parameters: $\mathbf{x}_{btg},\ \mathbf{x}_{bfg},\ \mathbf{r}_{lsp},\ \mathbf{u}_{fp}(t_0),\ m_{bi},\ m_{bf},\ T(t_0),\ \mathbf{T}_I^L(t_f^0),\ \mathbf{T}_L^P(t_0)$, and $\mathbf{T}_I^P$. It should be noted that the work presented here is a complimentary to the Klumpp's studies as it deals with trajectory design for powered descent with the following sequence: entry orbit to BP-CT arc to BP-quartic to AP-quartic to TDP. As the BP-CT arc solutions are of a closed-form, the entire design sequence is freed of the iterative procedures on the state vector and the target states. The interested readers can independently code and conduct simulations of this design sequence by using Eqs. (13.3), (13.35)–(13.47), (13.49) and (15.1). The BP-CT constraints, the inputs and outputs of the BP-targeting are given in Tables 15.2, 15.3 and 15.4.

Table 15.2 Input parameters for descent trajectory simulations

Entry orbit param. and LS	Position, velocity	Target ref. times	Time param. and angles
$r_p = 1735000$	$r_{xafg} = 30.0$	$T_{ai} = -165$	$\Delta t_{tdf} = 30.0$
$h_l = -2000.0$	$v_{xafg} = -1.0$	$T_{am} = -70.0$	$\Delta t_{cf} = 70.0$
$\theta_l = 0.0$	$r_{xamg} = 114.0$	$T_{af} = -25.0$	$\Delta t_e = 650.0$
$\phi_l = -89.90$	$v_{xamg} = -6.5$	$T_{bi} = -664.4$	$\Delta t_f^0 = 780.0$
$h_p = 15000$	$r_{xaig} = 2470.0$	$T_{bf} = -0.1$	$\gamma = 17.9$
$h_a = 100000$	$\lvert\mathbf{v}_{big}\rvert = 1665.0$	$\theta_0 = 16°$	

Table 15.3 Braking phase-quartic states (guidance)

State components	Initial state X Z	Final state X Z	Target state X Z
r	7048.538 –25357.156	2470.379 –7647.590	2466.596 –7632.343
v	–90.967 351.033	–37.874 152.614	–37.789 152.320
a	0.674 –2.734	0.843 –2.935	0.843 –2.935
j	0.002 –0.003	0.002 –0.003	0.002 –0.003
s	0.0 0.0	0.0 0.0	0.0 0.0

Table 15.4 Approach phase-quartic states (guidance)

State components	Initial state X Z	Final state X Z	Target state X Z
r	2470.379 –7647.590	30.0 –1.203	–90.914 21.807
v	–37.874 152.614	–1.0 0.481	–10.600 3.022
a	0.008 –1.976	–0.164 –0.096	–0.626 0.303
j	0.013 0.011	–0.016 0.016	–0.021 0.016
s	–0.0 0.0	0.0 0.0	–0.0 0.0

15.6 Analysis of Simulation Results

15.6.1 Qualitative Analysis

The simulations revealed many interesting features of the BP-quartic and AP-quartic. In particular, the AP-quartic trajectory is very sensitive to small changes in almost all targeting parameters and does not depend on the mass and engine parameters and the LS coordinates. The quartic formulation can work with a wide range of typical parameters, such as the states and the target referenced times, and can represent a trajectory with monotonically changing altitude, velocity and acceleration. Unlike the AP-quartic parameters, the BP-quartic trajectory depends on the mass, thrust vector, initial conditions of the jerk and snap vectors. In addition to the existing functional relationships between the quartic parameters, the only simulations with a number of sets of the trade parameters, such as γ, θ_{pbf} and p_1, can determine an appropriate set of the initial and target states that can satisfy the given requirements for the BP-quartic. Once these parameters are determined, the BP-quartic and AP-quartic can be connected easily by utilizing the continuity conditions. The BP-CT arc has also been shown to be sensitive to the changes in the mass and thrust parameters. Note that these solutions can generate a trajectory with decreasing velocity and slightly increasing altitude depending on the initial conditions. The analysis of the continuity conditions at the connection point between the BP-CT arc and BP-quartic have shown that the continuity of the state vector components can be satisfied yielding the appropriate duration of the BP-CT arc, the thrust percentage, desired altitude and velocity at the end point of the BP-quartic. The existence of the closed-form solutions for the BP-CT arc can increase the landing accuracy, which can be mainly determined by the distance between the $X_g - Z_g$ plane and the nominal $X_p - Z_p$ plane of the guidance and platform frames, respectively. Analysis of the simulation results, especially the Tables 15.3 and 15.4 and Figs. 15.2 and 15.3 have shown that the computation of the exact values of the constants given in Eqs. (13.28), (13.36)–(13.42), and the satisfaction of the continuity conditions using non-iterative procedures (except for the solution of Eq. (13.45) which is used only once when there is no re-targeting) have reduced the deviation of the $X_g - Z_g$ plane from the nominal $X_p - Z_p$ plane, and resulted in the landing in the close neighborhood of the desired LS (see Fig. 15.2).

15.6.2 Numerical Results

The numerical results of the simulations include the vectors computed by the targeting and guidance algorithms, the BP- and AP-state vectors at initial, final and target points of these phases. Given the specific times and targeting parameters, one can readily construct a corresponding nominal descent trajectory from the apolune of the entry orbit to the specified LS. In particular, a descent trajectory, designed by connecting the BP- and AP-quartics, can be used to design a nominal trajectory for any LS as the quartic formulations do not include the landing site coordinates. The numerical results of the simulations are given in Tables 15.3 and 15.4 and Figs. 15.3, 15.4, 15.5, 15.6, 15.7, 15.8, 15.9, and 15.10.

15.7 Trajectory Parameters

As it follows from the definition of the coordinate systems, the origin of the guidance coordinate system coincides with the predetermined LS at any time [1, 5] The LS is continuously changing its location with respect to the fixed platform coordinate system due to the Moon's rotation. The descent trajectory consisting of the BP-CT arc, BP-quartic and AP-quartic is presented in Fig. 15.3a, b. The simulation results consisting of the BP- targeting and AP- targeting parameters, that is the state vectors at the initial, final and target points of the BP and AP are presented in Tables 15.3 and 15.4. The entire simulated maneuver is starts at $t_x = 0$ and the BP-CT arc starts from the perilune of the elliptical entry orbit at 3409.9151 s, $(t_0 = 0)$ and lasts $t_1 = 439.52$ s. The following results for the position and velocity vectors have been obtained for the BP-CT arc in the guidance coordinate system:

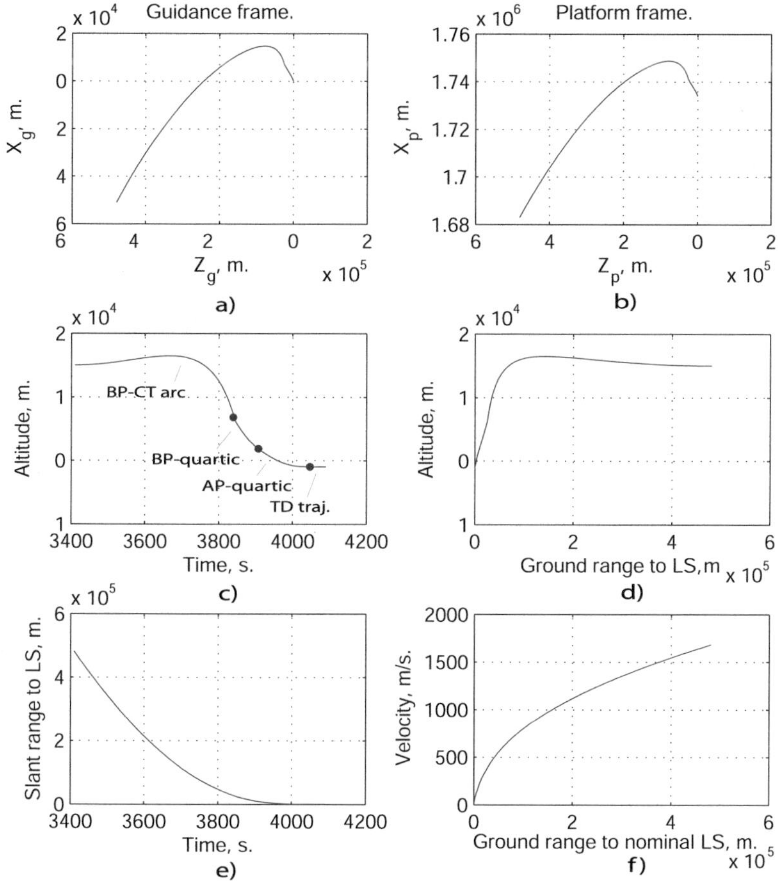

Fig. 15.3 Kinematic parameters of CT arc, BP, AP and TDP: **a** Trajectory in the guidance frame; **b** Trajectory in the platform frame; **c** Altitude versus time; **d** Altitude versus range to LS; **e** Slant range to LS versus time; **f** Velocity magnitude versus range to LS [1]

$$
\begin{gathered}
\mathbf{r}_{big}(-50981.231,\ 4.707,\ \ -479535.269), \\
\mathbf{v}_{big}(-460.605,\ 0.202,\ 1617.720), \\
\mathbf{r}_{bxg}(7048.538, 0.0, -25357.156), \\
\mathbf{v}_{bxg}(-90.967, 0.0, 351.033), \\
\mathbf{r}_{bfg}(2470.379, 0.0, -7647.590), \\
\mathbf{v}_{bfg}(-37.874, 0.0, 152.614).
\end{gathered}
$$

The components of $\mathbf{r}_{big}$ can be used to find the magnitude of this vector and the polar angle at the TCR point. The propulsion system parameters are $F = 90000\ N$, $p_1 = 0.9$ and $\beta = 25.4842$ kg/s. Equation (13.45) has been shown to be solv-

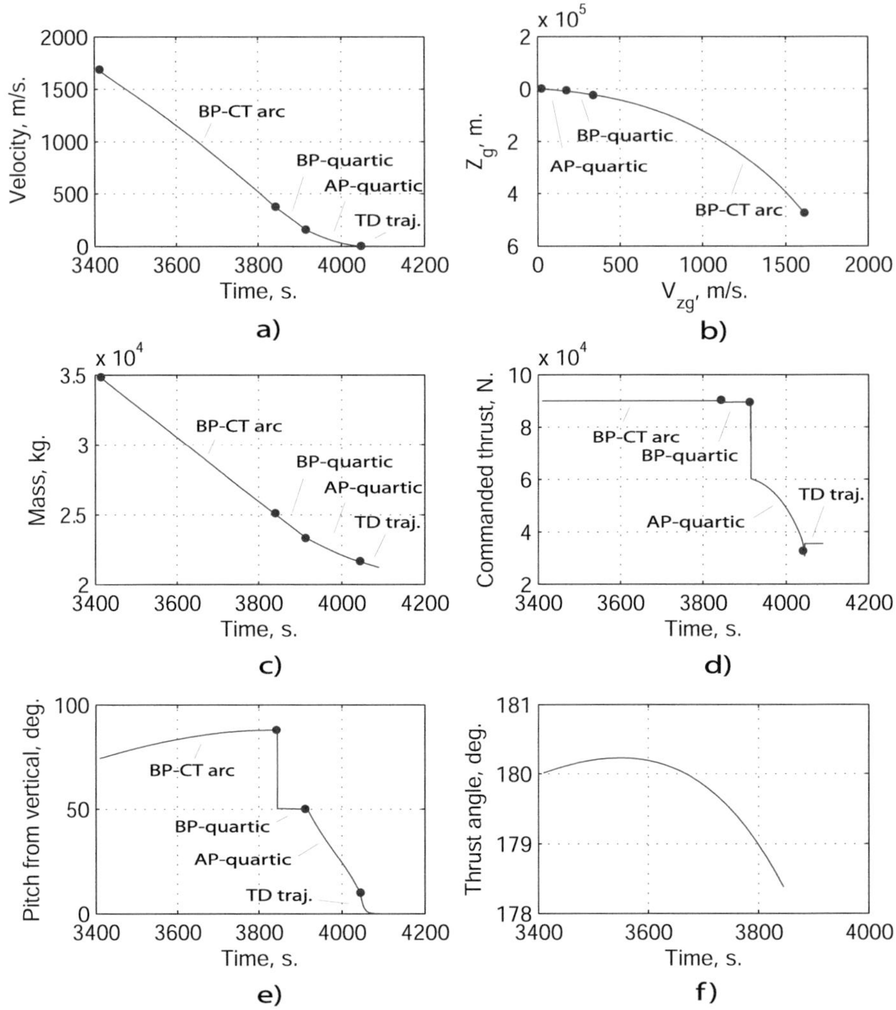

Fig. 15.4 Kinematic and dynamic parameters of CT arc, BP, AP and TDP: **a** Magnitude of velocity vector versus time; **b** Z- components of the position and velocity vectors; **c** Mass versus time; **d** Commanded thrust versus time; **e** Pitch from vertical versus time; **f** Thrust angle versus time on CT arc [1]

able for the families of the extremal trajectories when $p_1 \in [0.811, 0.974]$ and $t_1 \in [406.13\text{ s}, 487.72\text{ s}]$. It can be shown that the minimum fuel on the CT arc, that is the minimum J can be achieved by the lowest value of p_1, but not necessarily by the lowest value of t_1. For instance, if $p_1 = 0.88$ and $t_1 = 449.49$ s, then the corresponding fuel usage is $J = m_0 - m_1 = 10080.45$ kg. But if $p_1 = 0.94$ and $t_1 = 479.97$ s, then $J = 11497.87$ kg. The optimal values are $p_1 = 0.811$ and $t_1 = 487.72$ s have been shown to provide $J_{min} = 10080.18$ kg. For the extremal values $p_1 = 0.9$ and

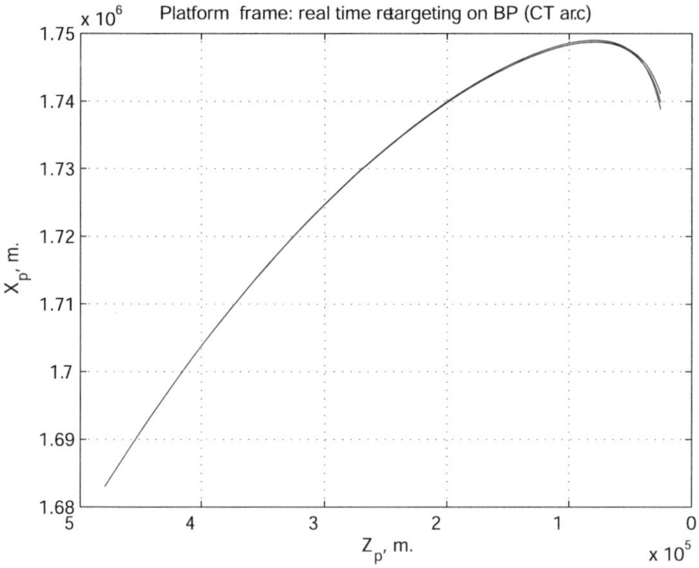

Fig. 15.5 BP re-targeting: trajectory in platform coordinates

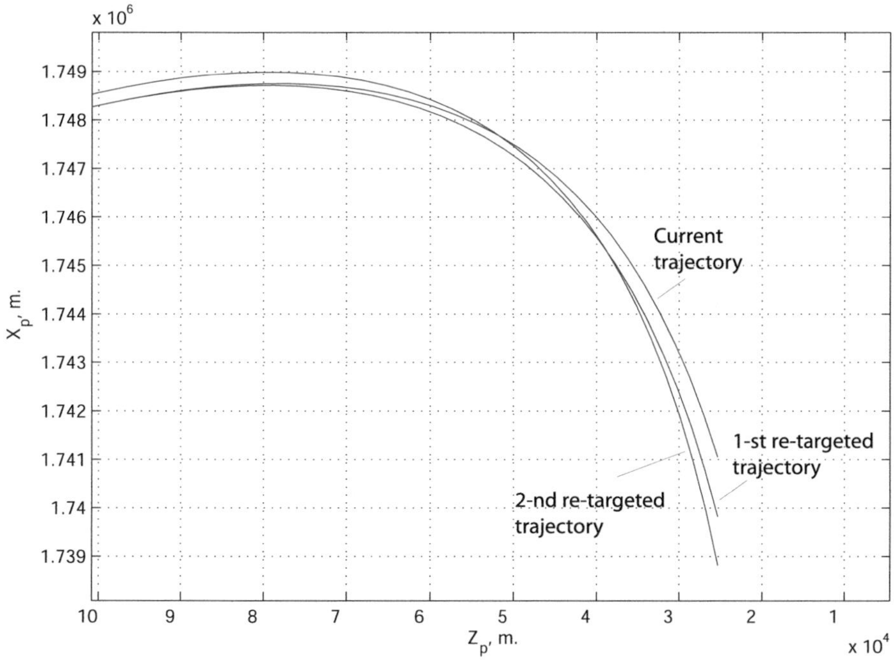

Fig. 15.6 BP re-targeting: trajectory zoom in

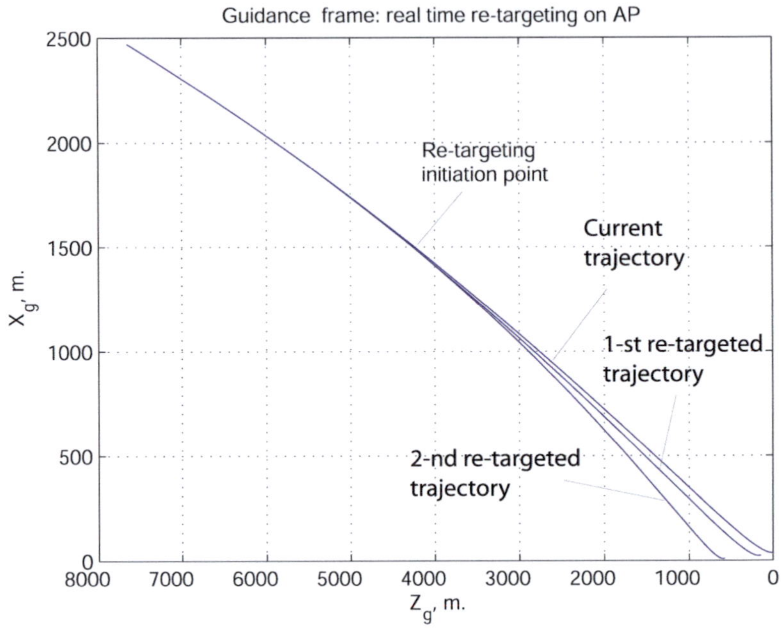

Fig. 15.7 AP re-targeting: trajectory in guidance coordinates

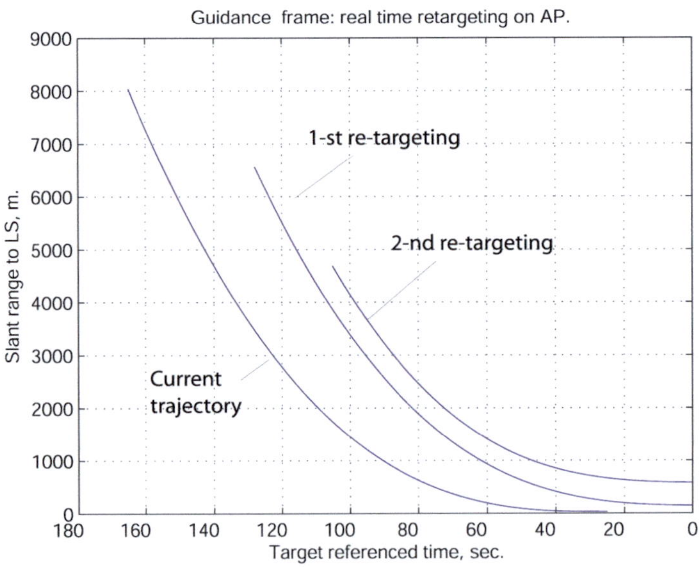

Fig. 15.8 AP re-targeting: slant range (LOC) versus target referenced time

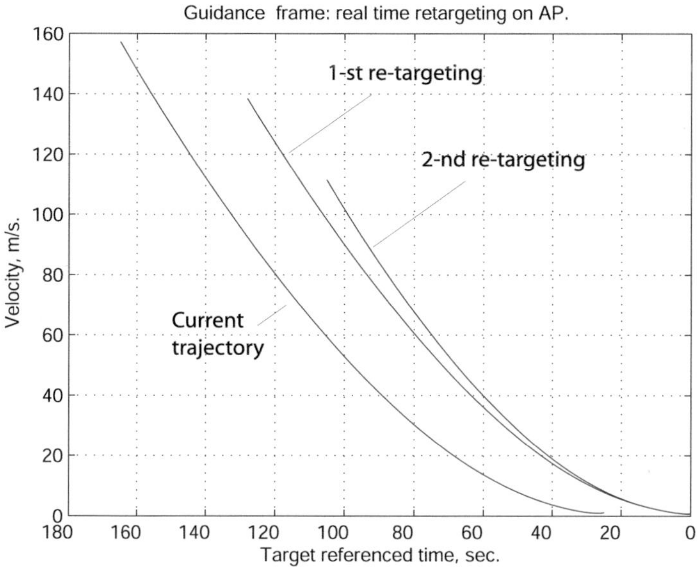

Fig. 15.9 AP re-targeting: magnitude of velocity versus target referenced time

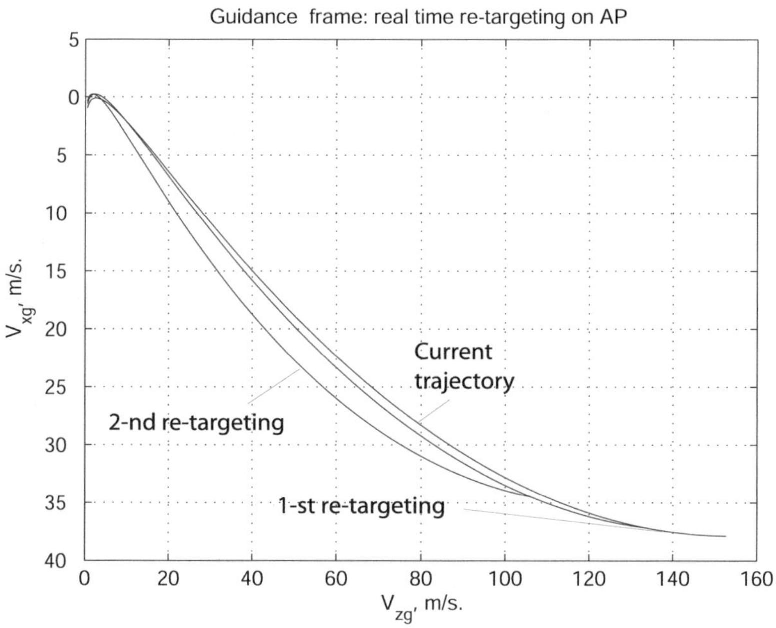

Fig. 15.10 AP re-targeting: velocity components

$t_1 = 439.52$ s used in the paper, one can show that $J = 10080.53$ kg. The powered descent and landing maneuver lasts 679.42 s and the landing occurs at 4088.8151s. The flight times on the BP-quartic and the AP-quartic are 70.0 s and 140.0 s respectively. The flight on the BP-quartic starts at 3844.51 s and ends 3914.71 s, and the TDP starts at 4044.515 s and lasts about 44.3 s. The trajectory in Fig. 15.3a has been generated using the guidance outputs whereas the plot (b) shows the nominal trajectory in the $X_p - Z_p$ plane of the platform frame. Also note for further analysis that the nominal parameters belong to the BP-quartic or AP-quartic and computed in the guidance coordinates. The relationships between the altitude, velocity slant range (line of site) and flight time are presented in Fig. 15.3c–f. The altitude is slightly increased in the beginning of the BP-CT arc due to the thrust direction and then gradually decreased until the landing (see Fig. 15.3c, d). Figures 15.3 and 15.4 show that the slant range and velocity are also gradually decreased in terms of time and ground range to the landing site (see Fig. 15.4a, b). The mass, commanded thrust, pitch from vertical and thrust angles are also presented in Fig. 15.4c–f. In addition to the plots of this figure, the final times of flight on each phase and the corresponding values of v_{gz} at each of these times are as follows:

$$
\begin{aligned}
&BP - CTarc : t_1 = 3844.51\text{ s},\ \ v_{gz}(t_1) = 351.03\text{ m/s};\\
&BP - quartic : t_2 = 3914.71\text{ s},\ \ v_{gz}(t_2) = 152.61\text{ m/s};\\
&AP - quartic : t_3 = 4044.51\text{ s},\ \ v_{gz}(t_3) = 0.48\text{ m/s};\\
&TDP : t_4 = 4088.82\text{ s},\ \ v_{gz}(t_3) = 0.0\text{ m/s}.
\end{aligned}
$$

As expected, the mass is linear function of time and the commanded thrust is constant on the BP-CT arc, and almost the same on the quartic trajectories due to the variable thrust commanded by the guidance algorithm (see Fig. 15.4c, d). Note that the commanded thrust is decreased slowly on the BP-quartic and rapidly on the AP-quartic. As commanded by the guidance algorithm, it stays constant on the TDP. The same changes can be observed in the behavior of the pitch angle from the vertical, except for the BP-CT arc, on which it gradually increases unlike the thrust angle given in the last plot. Note that the pitch angle is also rapidly decreased from about 9° to 0° on the TDP (see Fig. 15.4e). By definition, the thrust angle is the angle between the perpendicular to the position vector and the thrust vector (see Fig. 15.4f). Obviously, it is equal to 180° at the initial point of this arc, and then slowly increases and decreases for less than 2° by the end point of this arc. The simulations show that the reason for this change in the thrust angle is caused by the altitude and flight time.

15.8 Re-Targeted Trajectories

The simulations also included the re-targeting procedures. The unexpected internal or mechanical conditions, onboard terrain mapping, surface features, commands from communication sources and etc. may cause the on- board guidance program to change the trajectory and consequently perform a series of re-targeting procedures.

For illustration purposes, these procedures have been performed twice on the BP-CT arc and on the AP to demonstrate such capability provided by the closed-form solutions. For the BP, the results of these procedures are illustrated by the re-targeted trajectories in Fig. 15.5, and the last portions of the trajectories are shown in Fig. 15.6. For simulation purposes, only magnitudes of the position and velocity vectors are changed. The simulations have been performed two times using not the values computed according to the subsequent computations to satisfy the phase requirements, but the following values for these parameters:

$$r_1 = 1740000 \text{ m}, \quad v_1 = 340.0 \text{ m/s}; \quad r_2 = 1739000 \text{ m}, \quad v_2 = 310.0 \text{ m/s}.$$

These values changed the trajectory and resulted in considerably different landing positions (see Fig. 15.6). The same approach is used to compute the re-targeted trajectories twice on the AP. The simulation results are shown in Figs. 15.7, 15.8, 15.9 and 15.10. While on this phase, the altitude and altitude rate at the final point have changed:

$$r_{xafg}(1) = 22.0 \text{ m}, \quad r_{apfg}(1) = -145.0 \text{ m}, \quad v_{xafg}(1) = -0.6 \text{ m/s},$$

$$r_{xafg}(2) = 8.0 \text{ m}, \quad r_{apfg}(2) = -580.0 \text{ m}, \quad v_{xafg}(2) = -0.41 \text{ m/s}.$$

These parameters are not part of the input parameters. The first and the second re-targeted trajectories have landed about 145 m and 580 m away from the desired landing site or re-designated origin of the guidance frame (see Fig. 15.7). As a consequence of using the new parameters, the slant range has changed its behavior with respect to the target referenced times. Although these times reached zeros, the slant ranges to the landing site on the re-targeted trajectories have not been improved in terms of the target referenced time (see Fig. 15.8). This is qualitatively the same result illustrated in Fig. 15.7. The changes in the trajectory have also caused the changes in the target referenced times and in the components of the velocity vector in the guidance coordinates (see Figs. 15.9 and 15.10). The initiation of the first and the second re-targeting procedures can also be observed from Figs. 15.8 and 15.10.

15.9 Concluding Remarks

Development of the real-time targeting and re-targeting capabilities at any instant of powered descent, and the establishment of autonomous "communication" between targeting and guidance performances represent a novel methodology in advancing the state-of-the-art guidance technology. The initial and final conditions, and the general and mission-specific constraints can be satisfied by utilizing the arbitrary integration constants. The utility of the constants enhances the robustness of the algorithms to other maneuver constraints and mission-specific requirements. The future studies can be aimed at extension of the explicit on-board targeting and re-targeting solutions to other missions, further development of targeting and operational procedures, properties of quartic formulations aligned with the mission constraints and objectives.

References

1. Azimov DM, Bishop RH (2018) Integrated targeting and guidance for powered planetary descent. J Astronaut Sci 65:229–259. https://doi.org/10.1007/s40295-017-0126-4
2. Yang TL (1971) A targeting scheme for fuel optimal rocket trajectories with applications to the LM descent braking phase. Technical memorandum TM—71-2014-1, Bellcomm
3. Apollo 12 Mission Report (1970) Trajectory reconstruction and post-flight analysis. NASA-TM-X-70272, vol 1. Houston, Texas, p 113
4. Azimov DM, Bishop RH (2008) Reconstruction of Apollo targeting and guidance for powered lunar descent. New trends in astrodynamics and applications V, Milan, Italy
5. Klumpp AR (1971) Apollo lunar-descent guidance. Apollo guidance navigation control, R-695. Massachusetts Institute of Technology, MIT Instrumentation Laboratory Report, Cambridge, MA

Chapter 16
Landing Accuracy

The landing accuracy is another important parameter in the analysis of the descent and landing maneuver [1]. The exactness of the integration constants for the BP-CT arc and satisfaction of the continuity conditions without iterative procedures have been main factors in reducing the deviation of the $X_g - Z_g$ plane from the nominal $X_p - Z_p$ plane thereby providing the landing in the close neighborhood of the desired LS. Ideally, the r_g and r_p distances shown in Fig. 16.1 can be considered as two convenient measures of the landing accuracy. Here r_g is the distance from the actual LS and the LS re-designated for the last time before landing. r_p is the distance between the actual LS and the intersection of the X_p-axis of the $X_p - Z_p$ -plane and the lunar surface. However, due to the various reasons and/or from the realistic point of view, one would expect that the actual landing point would eventually deviate from the origin of the guidance frame (see Fig. 16.1). In this case, the location of the landing point with respect to these origins can serve as a measure of the landing accuracy. The computations have shown that the distances of the landing point from the origin of the guidance frame and from the intersection of the X_p-axis of the platform frame are $\Delta_{\rho g} = 4.629$ m and $\Delta_{\rho p} = 7.907$ m respectively. In the guidance frame, the downrange and cross-range of the landing point are 3.175 m and 0.042 m respectively. In the platform frame, these quantities are 6.501 m and 4.501 m respectively. It is also interesting to observe the time history of the difference between the lander's position vector components in both frames. As expected, this difference stays constant over the BP-CT arc and then changes rapidly afterwards until the TDP, where it again stays constant (see Fig. 16.2). This figure also shows that, in these simulations, unlike the alignment of the frames for the iteration purposes at the first guidance call in Apollo guidance, these frames were not aligned at the initial time. The time history of the rate of this difference is another helpful parameter in the analysis of the deviations of the frames from each other. Figure 16.3 shows that this rate is practically below 0.2 m/s for the entire trajectory and below 0.1 m/s for the BP- and AP-quartics. Note that the position and velocity vector components in both

© The Author(s), under exclusive license to Springer Nature Switzerland AG 2026

D. M. Azimov and R. H. Bishop, *Apollo-Class Targeting and Guidance for Lunar Powered Descent and Precision Landing*, Springer Aerospace Technology,
https://doi.org/10.1007/978-3-031-91088-3_16

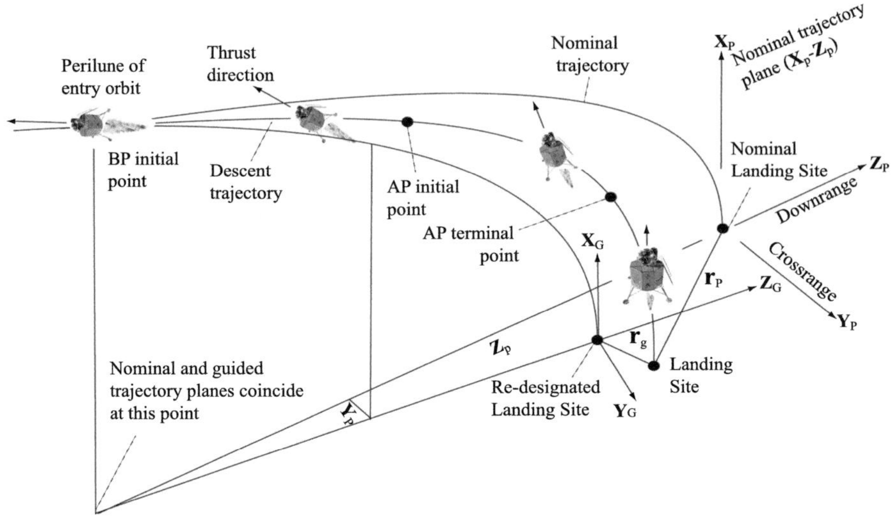

Fig. 16.1 Illustration of metrics of the landing accuracy in terms of platform and guidance frames

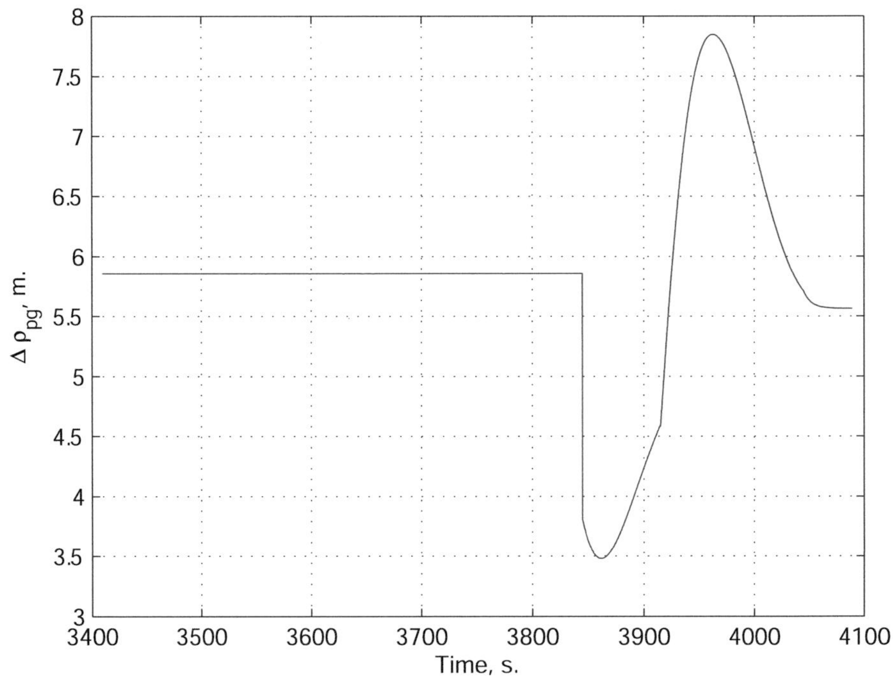

Fig. 16.2 Difference between the lander's position vector components in the platform and guidance frames

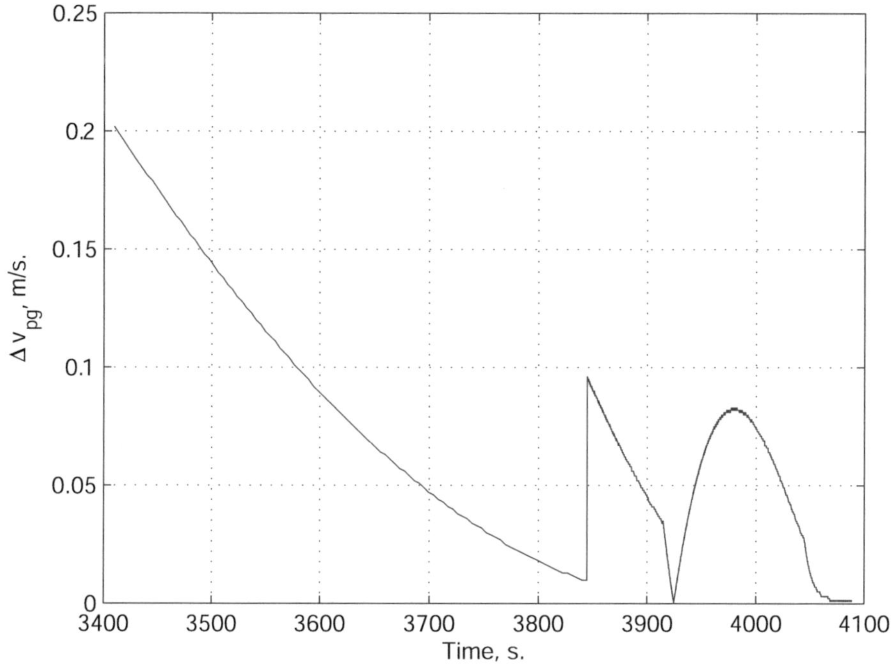

Fig. 16.3 Difference between the lander's velocity vector components in the platform and guidance frames

frames are the guidance functions and dependent on the commanded thrust angle and commanded coordinate transformations between these frames. The analysis of the simulation results confirm that the original Apollo guidance and targeting procedures can be extended and enhanced by utilizing the proposed closed-form solutions in the computations of the descent trajectory parameters and the re-targeting parameters along with the accuracy of landing. The time differences between the lander's position and velocity vector components in the platform and guidance frames are presented in Figs. 16.2 and 16.3.

Reference

1. Ennis WW (1971) An investigation of the landing accuracy detailed objective—case 310. Memorandum for file. B71 01028, Bellcomm, Inc.

Chapter 17
Conclusions

17.1 Reconstruction of Apollo Targeting and Guidance

In this monograph, the Apollo ground-based targeting, on-board ignition and guidance algorithms for the lunar descent and landing maneuvers have been described, reconstructed according to Apollo 11 and Apollo 12 missions' data and extended to include a landing maneuver at any point on the Moon. In particular, a landing maneuver at the Moon's South Pole has been discussed. The iterative procedures of the braking-phase targeting algorithm, analytical formulas for computation of the approach-phase targeting parameters, and the targeting, ignition and guidance programs are presented with the explicit content as much as possible.

The iterative procedures which take place in the braking-phase targeting algorithm, and certain aspects of both targeting and guidance procedures are presented in the most possible explicit forms. Some minor aspects of the original Apollo algorithms, such as the specific values of some coefficients, remain unknown.

The monograph describes the simulation results on Apollo targeting and guidance programs, and comparison analysis of these results with the trajectory reconstruction and post-flight analysis for Apollo Missions 11, 12 and a mission to the South Pole (Shackleton crater). The data compared by means of plots and the tables presented in the monograph demonstrate the closeness or similarity of the simulated trajectory data with the real mission post-flight data. The overall analysis of the results of the simulations show that the targeting and guidance programs used in this work can be used in the design of a soft and high-accuracy landing at any given landing site on the Moon surface, and other planetary bodies.

© The Author(s), under exclusive license to Springer Nature Switzerland AG 2026

D. M. Azimov and R. H. Bishop, *Apollo-Class Targeting and Guidance for Lunar Powered Descent and Precision Landing*, Springer Aerospace Technology,
https://doi.org/10.1007/978-3-031-91088-3_17

17.2 Extension and Enhancement of Apollo-Class Targeting and Guidance

Extension and enhancement of the Apollo guidance and targeting have been proposed through modifications to the braking phase of the powered lunar-descent and landing maneuver.

It has been shown that the application of closed-form solutions for constant thrust arc enables implementation of targeting algorithm on-board as many times as needed without engaging iterative procedures. Development of the real-time targeting and re-targeting capabilities at any instant of powered descent, and the establishment of autonomous "communication" between targeting and guidance performances represent a novel methodology in advancing the state-of-the-art guidance technology. The proposed algorithms for guidance and targeting enable the performance of various autonomous operations, such as mapping of the terrain features, hazard detection and avoidance, and re-designation of landing site. These algorithms have demonstrated the feasibility and technical likelihood of success of the future work.

The original Apollo ground-based targeting formulation, modification of the initial part of the BP and the simulations to explicitly compute the targeting parameters for the braking phase have been discussed. The part of the braking phase starting from the entry orbit and ending at the initial point of the throttle control recovery point was described by a closed-form extremal constant-thrust arc solution. Implementation of this solution avoids the repeated iterations on the state vector and target states and facilitates on-board execution of the braking-phase and approach-phase targeting and re-targeting procedures at any time during the descent thereby extending and enhancing the original Apollo guidance and targeting programs. The simulation results have shown that the shape of the finite thrust (maximum or constant) arc is highly dependent on the values of initial mass and the engine parameters, but not highly sensitive to changes in the initial entry orbit's parameters. Due to the sensitivity of the braking phase-quartic parameters, the constant-thrust arc can be connected to this quartic by selecting the specific values for certain parameters via the trade studies. The proposed integrated guidance and targeting design is simple, straightforward and free of the state vector and target state iterations. Understanding and proper interpretation of these algorithms are important in constructing the entire flight program for a soft landing on the Moon, and on any celestial body in general.

In all cases of simulations of the proposed analytical solutions for the braking phase, the targeting and guidance design parameters nearly satisfy the targeting conditions for both braking and approach phases, and the landing location have been shown to be very close (in the order of meters) to the origin of the guidance coordinate system, that is the nominal landing site. The specific values of the input parameters to these programs can be changed according to nominal landing design parameters.

The basic conclusion of the work presented is that the Apollo targeting and guidance problems can be solved entirely in explicit and closed form.

Regarding the future studies, it should be noted that in the context of the development of hazard avoidance technology and/or terrain selection purposes, the ground targeting program considered in this monograph should be freed from the iterative process, and re-designed to work on-board in order to provide real-time re-targeting capabilities. This can be accomplished by developing the explicit trajectory solutions for the braking phase with finite thrust and by utilizing the important properties of the quartic formulations of the trajectories near the planetary surface. These are the main topics for the future studies in the development of the targeting and guidance design for an autonomous landing with a hazard avoidance technology. The future studies may also include interpretation and explanation of the remaining on-board algorithms, including the attitude and throttle routines.

To summarize, the main results of the monograph are as follows:

- the simulation model utilizing the original Apollo targeting and guidance parameters have been developed;
- general quartic properties with purposes of using them in a maneuver trajectory design have been investigated;
- explicit on-board targeting by utilizing closed-form analytical thrust arc solutions and eliminating the uncertainties and iterations, which took place in the Apollo targeting and guidance algorithms has been presented and tested;
- Apollo ground-based targeting, on-board ignition and guidance algorithms for the lunar descent and landing maneuvers have been described and reconstructed according to Apollo 11 and Apollo 12 missions' data;
- Apollo-class real-time targeting, on-board ignition and on-board guidance algorithms for the lunar descent and landing have been enhanced;
- these algorithms have been extended thereby allowing us to perform a landing maneuver starting from the perilune of an arbitrary Moon orbit and ending at any point on the Moon.

The advantages of the proposed solutions and algorithms for Apollo-class real-time targeting and on-board guidance can also be extended to future cargo and robotic landing missions to planets and planetary moons. In particular, development of the proposed algorithms and capabilities for the robotic missions to Mars taking into account the atmospheric entry conditions and the conditions of supersonic and/or subsonic parachute deployment would be of a great interest.

MIX
Papier aus verantwortungsvollen Quellen
Paper from responsible sources
FSC® C105338
FSC
www.fsc.org

If you have any concerns about our products,
you can contact us on
ProductSafety@springernature.com

In case Publisher is established outside the EU,
the EU authorized representative is:
Springer Nature Customer Service Center GmbH
Europaplatz 3, 69115 Heidelberg, Germany

Printed by Libri Plureos GmbH
in Hamburg, Germany